Cambridge **Primary**

Ready to Go Lessons for Science

Step-by-step
lesson plans for
Cambridge Primary

Stage 2

Judith Amery

Series editor: Judith Amery

HODDER
EDUCATION
AN HACHETTE UK COMPANY

Note: Whilst every effort has been made to carefully check the instructions for practical work described in this book, schools should conduct their own risk assessments in accordance with local health and safety requirements.

Every effort has been made to trace all copyright holders, but if any have been inadvertently overlooked the Publishers will be pleased to make the necessary arrangements at the first opportunity.

Although every effort has been made to ensure that website addresses are correct at time of going to press, Hodder Education cannot be held responsible for the content of any website mentioned in this book. It is sometimes possible to find a relocated web page by typing in the address of the home page for a website in the URL window of your browser. Websites included in this text have not been reviewed as part of the Cambridge endorsement process.

Hachette UK's policy is to use papers that are natural, renewable and recyclable products and made from wood grown in sustainable forests. The logging and manufacturing processes are expected to conform to the environmental regulations of the country of origin.

Orders: please contact Bookpoint Ltd, 130 Milton Park, Abingdon, Oxon OX14 4SB. Telephone: (44) 01235 827720. Fax: (44) 01235 400454. Lines are open 9.00–5.00, Monday to Saturday, with a 24-hour message answering service. Visit our website at www.hoddereducation.com.

© Judith Amery 2013
First published in 2013 by
Hodder Education,
An Hachette UK Company
338 Euston Road
London NW1 3BH

Impression number   5 4 3 2 1
Year   2017 2016 2015 2014 2013

Cover illustration by Peter Lubach
Illustrations by Planman Technologies
Typeset in ITC Stone Serif Medium 10/12.5 by Planman Technologies
Printed in Great Britain by CPI Group UK (Ltd), Croydon, CR0 4YY

A catalogue record for this title is available from the British Library.

ISBN: 978 1444 177831

Contents

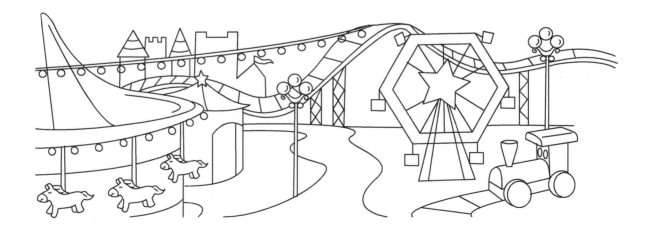

Introduction

About the series

Ready to Go Lessons is a series of photocopiable resource books providing creative teaching strategies for primary teachers. These books support the revised Cambridge Primary curriculum frameworks for English, Mathematics and Science at Stages 1–6 (ages 5–11). They have been written by experienced primary teachers to reflect the different teaching approaches recommended in the Cambridge Primary Teacher Guides. The books contain lesson plans and photocopiable support materials, with a wide range of activities and appropriate ideas for assessment and differentiation. As the books are intended for international schools we have taken care to ensure that they are culturally sensitive.

Cambridge Primary

The Cambridge Primary curriculum frameworks show schools how to develop the learners' knowledge, skills and understanding in English, Mathematics and Science. They provide a secure foundation in preparation for the Cambridge Secondary 1 (lower secondary) curriculum. The ideas in this book can also be easily incorporated into existing curriculum frameworks already in your school.

How to use this book

This book covers each of the units of the scheme of work for Science at Stage 2. It can be worked through systematically (as all the learning objectives are covered), or used to support areas where you feel you need more ideas. It is not prescriptive – it gives ideas and suggestions for you to incorporate into your own teaching as you see fit.

Each step-by-step lesson plan shows you the learning objectives you will cover, the resources you will need and how to deliver the lesson.

Each lesson includes a Starter activity, Main activities and a Plenary that draws the lesson to a close and recaps the learning objectives. Success criteria are provided in the form of questions to help you assess the learners' level of understanding. The 'Differentiation' section provides support for the less-able learners and extension ideas for the more able.

For each lesson plan there is at least one supporting photocopiable activity page. At the end of each unit there are also suggestions for assessment activities. Answers to activities can be found at www.hoddereducation.com/checkpointextras.

Learning objectives

The *Science Curriculum Framework* provides a set of learning objectives for each stage. At the start of each lesson you need to re-phrase the learning objectives into child-friendly language so that you can share them with the learners at the outset. It sometimes helps to express them as *We are learning to / about ...* statements. This really does help the learners to focus on the lesson's outcomes. For example: 'Know that water is taken in through the roots and transported through the stem' (Stage 3) could be introduced to the learners at the start of the lesson as: *We are learning about the journey water takes through a plant.* To avoid unnecessary repetition we have not included such statements at the start of each lesson plan but it is understood that the teacher would do this.

The overview chart on pages 6–7 shows you how the learning objectives are covered in the lessons in this book.

Time commitment

Teachers should be aware that the recommended time commitment for Science at Stages 1 and 2 is an hour to one and a half hours per week. This could be as a whole afternoon or two or three shorter sessions, depending on timetabling arrangements in your school. The recommended time commitment at Stages 3 to 6 is at least two hours per week. This provides ample time to carry out practical work. Again, it can be timetabled as one long or several shorter sessions. We have, however, provided the same number of lesson plans for you for all six stages to provide choice and variety. Please select the most appropriate lessons for your class for Stage 2 to suit the amount of time available to you.

Success criteria

These are the measures that the teacher and, eventually, the learner will be able to use to assess the outcome of the learning that has taken place in each lesson. They are included as a series of questions, which will help you as teacher to assess the learners' understanding of the skills and knowledge covered in the lesson.

Scientific enquiry skills

Science teaching is concerned with more than just the learning of scientific facts. Scientific enquiry skills are also **essential**.

The activities in these books will show you how to incorporate scientific enquiry skills in order to link practical skills alongside thinking skills using the Cambridge Primary Science Programme. Scientific enquiry is embedded in the curriculum in the Biology, Chemistry and Physics strands. The skills of scientific enquiry are on-going in each stage and between stages. These skills need to be used regularly, in familiar and new contexts, in order for the learners to become young scientists who are capable of questioning, reasoning and finding answers through scientific investigation. Every lesson in this book has links to at least one scientific enquiry learning objective.

The key to successful scientific enquiry teaching lies in providing the learners with opportunities to learn by doing, that is, through **active learning**.

Formative assessment

Formative assessment is on-going assessment that occurs in every lesson and informs the teacher and learners of the progress they are making, linked to the success criteria. The types of questions to ask that will support teachers in making formative assessments have been incorporated into each lesson in the 'Success criteria' sections.

One of the advantages of formative assessment is that any problems that arise during the lesson can be responded to immediately. Formative assessment influences the next steps in learning and may influence changes in planning and / or delivery for subsequent lessons.

Summative assessment

Summative assessment is essential at the end of each unit of work to assess exactly what the learners know, understand and can do. The assessment sections at the end of each chapter are designed to provide you with a variety of opportunities to check the learners' understanding of the unit. These activities can include specific questions for teachers to ask, activities for the learners to carry out (independently, in pairs or in groups) or written assessment.

The information gained from both the formative and summative assessment ideas can then be used to inform future planning in order to close any gaps in the learners' understanding as recommended by *Assessment for Learning* (AFL).

Safety

All the lessons in this book have been written with safety in mind. However, please ensure that you are aware of and conform to any national, regional or school regulations for safety as you conduct any of the activities in this book. Always be aware of skin and food allergies or intolerances and obtain parental consent for the learners to participate in tasting activities. If necessary, make sure that you undertake a risk assessment of potential hazards before undertaking activities. It is important to ensure that the learners are aware of safety considerations when carrying out practical activities.

Appropriate use of ICT

At the planning stage teachers need to consider how the use of ICT in a lesson will enhance the learning process. Ensure that the ICT resources you use support and promote the learners' understanding of the learning objectives. Activities included in this book have been designed to be carried out without the need for state-of-the-art ICT facilities. Suggestions have also been included for schools with internet access and / or the use of interactive whiteboards. This is in order to cater for most teachers' needs.

In these lessons the author sometimes asks for the teacher to display an enlarged version of the photocopiable page at the front of the class. We have not specified whether this should be using an overhead projector, interactive whiteboard or flipchart, as schools will have different resources available to them.

We hope that using these resources will give you confidence and creative ideas in delivering the Cambridge Primary curriculum framework.

Judith Amery, Series Editor

Overview chart

Term 3

Sources of light

Learning objectives

- Make comparisons. (2Eo6)
- Identify different light sources, including the Sun. (2Pl1)

Resources

Flipchart and markers or interactive whiteboard; photocopiable pages 9 and 10.

Starter

- Ask the learners to discuss with talk partners: *Where does light come from?*
- Listen to the learners' responses and list their suggestions on the flipchart or interactive whiteboard.
- Explain that our main source of light on Earth is the Sun. Describe it to the learners as a giant ball of burning gas in space. Tell them that as it burns, it glows and gives out light that travels to Earth to make daylight for us.

Main activities

- Look at the list of learners' suggested sources of light. Ask: *Which of these are natural sources of light?* (For example the Sun, stars, lightning, fire, fireflies, some deep-sea fish.) Discuss as a class how each of these things produces light:
  - the Sun – burning gas glowing
  - stars – balls of burning gas that can be smaller or even larger than the Sun – but tend to look smaller as they are further away
  - lightning – a bright flash of electricity made by a thunderstorm
  - fire – flames made from burning wood or fuel
  - fireflies – special cells in their bodies (abdomen) can make light
  - deep-sea fish – also have special cells in their bodies that help them to make light.
- Next discuss man-made (artificial) sources of light, choosing examples from the class list (for example oil lamps, candles, torches, electric lamps). Ask the learners where the light comes from in each of these artificial sources, that is, lamp oil for oil lamps, wax for candles, cells (batteries) for torches and electricity for lamps.

- Give out photocopiable page 9 (to the learners who need support) or photocopiable page 10 (to all the other learners) and explain that they need to identify and classify different sources of light in the pictures shown.

Plenary

- Ask the learners to share their responses to photocopiable pages 9 and 10.
- Discuss each response in turn and classify it as a natural or artificial source of light.
- Ask the learners to show or find sources of light inside the classroom.
- Ask them to bring in examples of light sources to make a classroom display.

Success criteria

Ask the learners:

- What is the main source of light for us on Earth? (The Sun.)
- Can you name a natural source of light?
- Does a torch give natural or artificial light? (Artificial.)
- What is the source of the light from a torch? (The cell [battery].)

Ideas for differentiation

Support: Assist these learners in copying the words from the list correctly on photocopiable page 9.

Extension: Ask these learners to look through brochures and / or magazines and to cut out pictures showing different sources of light. Make them into a collage for display.

Name: _____

Sources of light

1. Use these words to name the pictures below that show different sources of light.

| candle | fire | firefly | lamp | Sun | torch |

2. Now name three more sources of light:

a) _____

b) _____

c) _____

Name: _____

Sources of light

1. Draw and label four different sources of light in the boxes below. Use the class list to help you.

_____	_____
_____	_____

2. Complete the table to show which of the light sources you have chosen are natural and which are artificial.
   One example has been done for you.

Light source	Natural?	Artificial?
The Sun	✓	

What is light?

- Collect evidence by making observations when trying to answer a science question. (2Ep1)
- Identify different light sources including the Sun. (2Pl1)

Photocopiable page 12.

Starter

- Have a quiz about light. Allow the learners to work as teams on their tables, or split them into two or several teams to take the quiz. Ask the following questions in turn. Award a point for a correct answer. Do not subtract points for incorrect answers. The winning team is the team with the most points.
  Quiz questions:
  - What do our eyes help us to do? (See.)
  - What is the main source of light for us on Earth? (The Sun.)
  - Which other senses would we need to use if we could not see? (Taste, smell, touch, hearing – award one point for each correct answer, repeat the question to each team until all four correct responses have been given.)
  - What does sunlight help green plants to do? (Make their food.)
  - When does daytime start? (At sunrise – when the Sun rises in the morning.)
  - What do we call the time of day when daytime ends? (Sunset.)

Main activities

- Explain that light is usually given out by things that get very hot and glow, so giving out light. The Sun is so hot and glows so brightly that it gives us daylight on Earth every day. A candle flame or a burning fire only gives out a small amount of light.
- Explain that moonlight is not really light at all. The Sun shines on the moon and the moon reflects the Sun's light back to us on Earth.

- Ask the learners to discuss with talk partners how the Sun's light is useful to us during the day.
- Listen to the learners' responses and ideas. Give each pair of learners an opportunity to contribute if they so wish.
- Give out photocopiable page 12 and explain that it shows some sources of light. Some were mentioned in the previous lesson and some will be new to the learners. This time they have to name the object and / or its light source.

Plenary

- Discuss the learners' responses to photocopiable page 12. Explain that an oil lamp burns lamp oil to produce a flame and a fire burns wood.
- The learners should be able to give numerous examples of objects that have electricity as their light source.

Ask the learners:

- Why is the Sun so important to us on Earth? How do you know this?
- What do green plants use sunlight for?
- Can you name something that has electricity as its source of light?
- What do we call the start of the day?
- What happens at sunset?

Support: Assist these learners if they are a quiz team in themselves. Or, put them in mixed-ability teams to do the quiz. Discuss their suggestions of objects that have electricity as their source of light.

Extension: Ask these learners to design and display a poster about different everyday sources of light.

Name: _____

More light sources

1. Complete the table to show objects and their source of light.

2. Draw and write the name of an object for section 2.

3. Write the name of the light source for objects 1 and 3.

Object	Light source
1. oil lamp	_____
2. _____	electricity
3. fire	_____

Cambridge Primary: Ready to Go Lessons for Science Stage 2 © Hodder & Stoughton Ltd 2013

What is dark?

Learning objectives

- Use first-hand experience. (2Ep2)
- Review and explain what happened. (2Eo9)
- Know that darkness is the absence of light. (2Pl2)

Resources

A globe; a torch; a toy figure of a person; sticky tack; photocopiable pages 14 and 15; empty boxes with lids, e.g. shoeboxes; black paper or paint and paintbrushes; aprons; glue; scissors; torches and cells (batteries); shiny objects, e.g. keys, aluminium foil, jewellery, metal spoons.

Starter

- Recap from the previous lesson that the Sun gives us daylight on Earth. Tell the learners that the Sun is so big and so hot that on Earth we are just the right distance in space away from it. Remind them that the Sun gives us enough light and heat to survive and live well here on Earth.

- Attach the toy figure to your country on the globe. Ask a learner to pretend to be the Sun by shining the torch on the globe as you rotate it. Demonstrate the Earth rotating on its axis.

- Stop the globe at various times. Ask: *What time would it be in our country now?*

Main activities

- Explain that the Earth spins once a day; it is daytime when your country is facing the Sun and night-time when your country is facing away from the Sun. Repeat the Starter activity each time you explain this – this is a difficult concept for young learners to understand.

- Ask the learners to discuss with talk partners what they do at night-time.

- Listen and respond to the learners' answers (for example go to sleep, stay indoors, stop work). Discuss any jobs that people do at night-time.

- Give out photocopiable pages 14 and 15 and explain that the learners will test to find out what it is like in the dark. Explain that they are going to make a dark box and use it to discover how well or badly they are able to see things in it. Demonstrate how to follow the instructions on photocopiable page 14 to make such a box. Read the instructions out one by one, or ask the learners to read them out for you to follow. Alternatively, as you read the instructions aloud, the learners should follow them to make a box. Check that all the torches available are in good working order, and replace the cells (batteries) if necessary.

Plenary

- Discuss the learners' findings – do they all agree?

- Explain that it is dark because there is no light shining. Dark is the absence of light. At night-time (sunset to sunrise), it is a good time for us to sleep.

Success criteria

Ask the learners:

- Why do we have night?
- When does night-time begin?
- When does it end?
- What kinds of lights do we use at night-time?
- What is dark?

Ideas for differentiation

Support: Assist these learners with making their box. Help them to complete their ideas in writing on photocopiable page 15.

Extension: Ask these learners to make a list of different lights that are used at night-time.

Name: _____

In the dark

What to do

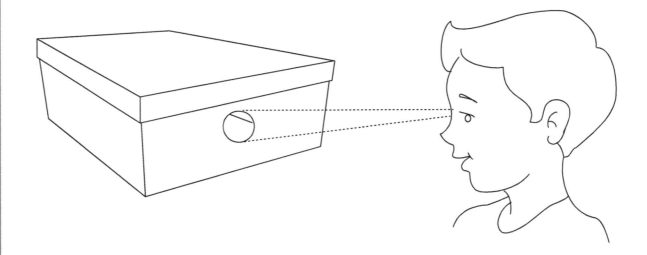

- Paint or cover the inside of the box with black paper or paint.

- Make a small hole in one side of the box.

- Put the lid on the box and look through the hole.

1. Can you see anything? yes / no

2. Put a shiny object inside the box and put the lid back on.
    Look through the hole again. Can you see the object? yes / no

3. Do this with another two different shiny objects.
    Can you see the object?

    a) Object: _____ yes / no

    b) Object: _____ yes / no

4. Switch on the torch and put it in the box.

5. Use the 'What happened' page to record what you did.

Name: _____

What happened

1. Draw a picture of your box with the lid on.

2. Draw where you put the hole.

3. The objects I used were:

   a) _____

   b) _____

   c) _____

4. What happened when I put the torch in the box:

   _____

   _____

   _____

Light and dark

Learning objectives

- Talk about risks and how to avoid danger. (2Eo2)
- Identify different light sources, including the Sun. (2Pl1)
- Know that darkness is the absence of light. (2Pl2)

Resources

Flipchart and markers or interactive whiteboard; blindfolds; PE equipment, e.g. cones; a large space; photocopiable page 17.

Starter

- Recap that when it gets dark outside, we turn on electric lights. Ask: *Why do we do this?* (To help us to see.)
- Make a class list of all the different kinds of electric lights that the learners can think of. Display it clearly for them all to see and be able to refer to (for example lamps, street lights, carnival lights, street signs, torches).

Main activities

- Ask the learners to discuss with talk partners: *Which places use electric light at night?*
- Listen to and discuss their responses (for example homes, streets, factories, cinemas, airports, railway stations, shops).
- Give out photocopiable page 17 for the learners to complete with the correct names.
- Read or tell a story about light and dark, such as *The Owl Who Was Afraid of the Dark* by Jill Tomlinson (Egmont Books Ltd). This can be found by searching on YouTube by title and author. There are lots of stories available on this theme – choose a favourite one that is familiar to the learners if possible. This will encourage them to talk more freely about their feelings of being in the dark.
- Ask them to tell you some places that are dark (for example caves, tunnels, mines, deep sea). Ask: *How does being in the dark make you feel?*

- Talk about their experiences of having been to or having seen such places (if any). Remember that many young learners may still be afraid of the dark.
- Explain that these places are dark because sunlight cannot reach them.
- Play a game – finding your way around in the dark. In pairs, if the learners consent to do so, give one learner a blindfold to put on. With the help of their partner, go around a simple course, laid out using PE equipment. Take it in turns.
- Discuss how this experience made them feel. Ask: *Was the help of a friend important?*

Plenary

- Recap the importance of being able to see because of light.
- Think about places the learners are familiar with that use light at night.
- Explain darkness as the absence of light and how people prefer to be in the light.

Success criteria

Ask the learners:

- Why can't we see (or see so well) in the dark? (Because there is no / little light.)
- Name some of the kinds of lights you have at home at night-time.
- Do you prefer the light or the dark? Why?

Ideas for differentiation

Support: Support these learners in discussion and give help when going around the obstacle course, if necessary.

Extension: Ask these learners to find out about an animal that lives in the dark (for example moles, some deep-sea fish or creatures, earthworms).

Name: _____

Turning on the lights

Electric lights help us to see when it is dark.

1. Look at these pictures with lights in them.

2. Use these words to name the lights in each picture
    (you may have to use some words more than once).

floodlights   headlights   lamp   sign   torch   traffic lights

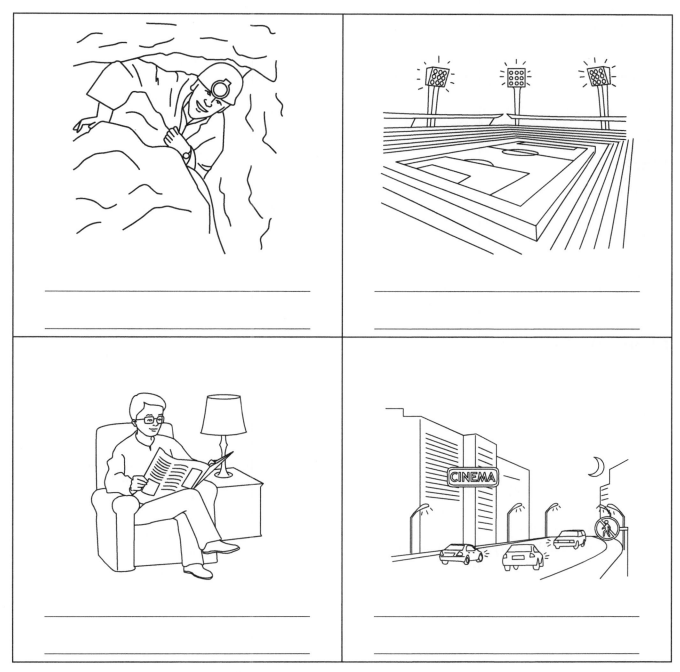

_____

_____

_____

_____

_____

_____

Light and dark stories

Learning objectives

- Use simple information sources. (2Ep3)
- Identify different light sources, including the Sun. (2Pl1)
- Know that darkness is the absence of light. (2Pl2)

Resources

Blank flashcards; marker; a selection of pictures (from books or the internet) of light and dark settings, e.g. a forest, a cave, under the sea, an amusement park, a thunderstorm; flipchart and markers or interactive whiteboard or computer; photocopiable pages 19 and 20.

Starter

- Ask the learners to think with talk partners of as many words as they can that describe dark or light things (for example 'dark', 'dim', 'gloomy', 'shady', 'light', 'sparkling', 'twinkling', 'flashing', 'gleam', 'glow', 'dazzle').
- Write the learners' suggestions on the blank flashcards and display them clearly for all to see and refer to during the lesson.

Main activities

- Show the learners a picture of one of the settings available.
- Model and write a story with the learners to go with the picture, using some of the words generated in the Starter activity, for example *We're Going on a Bear Hunt* by Michael Rosen (Walker Books).
- Invite the learners to 'have a go' at writing some sentences on the flipchart, interactive whiteboard or computer for you.
- Read and re-read the story aloud, modelling how to make appropriate corrections throughout. Discuss word choices: *Is there a better word we could use here? Can we change the order of words in a sentence to make it sound better?*

- Give out photocopiable pages 19 and 20. Explain to the learners that they have to choose one of the pictures to write a story about. They need to use as many good, descriptive words as they can. They can use the class word list to help them.
- Tell them that their finished stories will be bound into a book for the class library, so you are looking for their best work.

Plenary

- Look at each of the picture stimuli from photocopiable page 19 in turn.
- Ask the learners to give you a descriptive sentence of the setting.
- Listen to some of the stories or parts of stories the learners have written.

Success criteria

Ask the learners:

- Choose three words from the flashcards to describe a forest.
- Give me three good describing words for a thunderstorm.
- Read me a sentence with lots of words describing light.
- What setting did you choose for your story? Why?
- Show me a good opening sentence you have written.

Ideas for differentiation

Support: Work with these learners to support their observations of the picture they choose, or select the picture for them and work as a small group on one story together.

Extension: Ask these learners to write their stories for the learners in Stage 1. Make these into a separate book as a gift to that class.

Light and dark stories

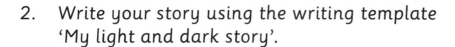

1. Choose one of the pictures below to write a story about.

2. Write your story using the writing template 'My light and dark story'.

3. Use the words you have thought of in class in your story.

In the forest

At the theme park

The thunderstorm

Name: _____

My light and dark story

Write your story. Underline all the good describing words in it. Draw a picture from your story in the box below.

All the stories will be made into a class book.

Title: _____

_____

_____

_____

_____

_____

_____

_____

Cambridge Primary: Ready to Go Lessons for Science Stage 2 © Hodder & Stoughton Ltd 2013

Celebrate with light

Learning objectives

- Use first-hand experience. (2Ep2)
- Make comparisons. (2Eo6)
- Identify different light sources, including the Sun. (2Pl1)

Resources

A music or sound system; music from light festivals; pictures (from books or the internet) or artefacts of objects associated with festivals of light, e.g. Hanukkah, Divali, Bonfire Night, Christmas (artefacts could include candles, Christmas lights, oil lamps); photocopiable page 22.

Starter

- Sing a celebration song or songs or listen to songs associated with a festival or celebration that includes light, for example Divali, birthdays, Christmas, Hannukah, Bonfire Night. Alternatively, listen to some music associated with a particular festival of light.
- Talk about some festivals that include lights. Ask: *When are they celebrated? Which countries celebrate in this way?*
- Ask the learners to tell the rest of the class about any of their festival experiences.
- Ask them to bring in photographs of their festival experiences, if permitted.

Main activities

- Show the learners some of the pictures or artefacts associated with particular festivals.
- Ask the learners to discuss with talk partners which festival each artefact is part of, and identify what they know already about that particular celebration.
- Listen to their responses as a class.
- Discuss the celebrations in turn and describe how the artefacts are used as part of the ceremony.
- Re-enact parts of the celebrations, using the artefacts (if available), or make simple props for the learners to use.

- Read a story by candlelight or reading lamp. This could be a traditional story or a story that has light and dark in it. It could be the same story as used in the previous lesson. Discuss the learners' feelings about light and dark.
- Give out photocopiable page 22 and explain to the learners that they need to look at the pictures and decide which festival each picture shows. They then need to copy the name of the festival underneath each picture and colour in the pictures.

Plenary

- Compare the similarities and differences between the different festivals, for example all involve lights, they are held at similar (dark) times of year.
- Ask some of the learners to show their completed photocopiable page 22.

Success criteria

Ask the learners:

- What is the light source for a Divali lamp? (Lamp oil.)
- What makes Christmas lights light up? (Electricity or cell [battery] power.)
- What do each of the festivals we have talked about have in common? (They use light.)
- What time of year are these festivals celebrated? (Dark season.)
- Which is your favourite festival and why?

Ideas for differentiation

Support: Help these learners when using props or artefacts to re-enact parts of ceremonies. Ask open-ended questions to help them to focus their thoughts.

Extension: Ask these learners to make a poster for a light festival of their choice to be displayed in the classroom at festival time.

Name: _____

Festivals of light

1. Look at the lights in these pictures.

2. Choose which festival is being celebrated from the list below.

| Bonfire Night | Christmas | Divali | Hanukkah |

3. Label and colour the pictures.

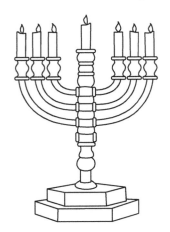

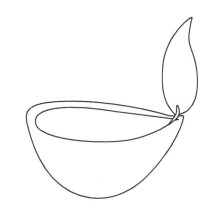

_____

_____

_____

_____

Cambridge Primary: Ready to Go Lessons for Science Stage 2 © Hodder & Stoughton Ltd 2013

Shadows

- Collect evidence by making observations when trying to answer a science question. (2Ep1)
- Use first-hand experience. (2Ep2)
- Be able to identify shadows. (2Pl3)

A large outside space in the Sun; photocopiable page 24.

Starter

- Go outside into a sunny area. Ask the learners to stand with their backs to the Sun.
- Talk about the dangers of looking directly at the Sun and discourage this. (It can damage your eyes – wear sunglasses and / or a hat for protection.)
- Look at the shadows the learners' bodies make on the ground.
- Ask: *Can you stand on your own shadow? Try it and see!*

Main activities

- Play a game of chasing shadows. Select a few learners to be 'it'. On the signal, the learners must run around inside a defined area in the sunny part of the outdoor space. A learner is caught out by an 'it' person if the 'it' person stands on a learner's shadow. That learner then has to stand still. The learners can be released by another learner standing on their shadow. The game continues until everyone apart from the 'it' people are still. The last person to be still is the winner.
- Ask the learners to make, for example, the biggest shadow that they can, the smallest shadow, the tallest shadow, the narrowest shadow, and so on. Discuss and observe exactly how they did this.

- Ask them to observe their own and the other learners' shadows when you instruct them to hop, skip, jump, and so on. Spend some time on these activities to ensure that all the learners have had the opportunity to see and make a variety of shadows.
- Give out photocopiable page 24 and explain to the learners that they have to look around to find some different shadows and draw them. They also have to try to lose their shadow and be able to write a sentence about how they did it.

Plenary

- Ask the learners to show the rest of the class how they managed to escape from their own shadows. (Expect some very creative responses!)
- Explain that shadows are made when light is being blocked. The light can be from the Sun or any other light source. Their bodies made shadows because the light from the Sun was shining on them, but could not go through their bodies.

Ask the learners:

- What makes a shadow? (The light is blocked.)
- How did you make your shadow disappear?
- How can you make a big shadow?
- How can you make a small shadow?
- Has your shadow changed in size since the start of the lesson?

Support: Work in a small group with these learners to try out all their ideas of how they might lose their shadows.

Extension: Ask these learners to find the biggest and smallest shadows they can. Write or draw them on the back of photocopiable page 24.

Name: _____

Shadows

1. Find some shadows outside.

2. Draw what you see.

3. Can you make your shadow disappear? yes / no

4. I made my shadow disappear by:

_____

_____

 Cambridge Primary: Ready to Go Lessons for Science Stage 2 © Hodder & Stoughton Ltd 2013

Making shadows 1

Learning objectives

● Make suggestions for collecting evidence. (2Eo1)

● Use a variety of ways to tell others what happened. (2Eo5)

● Be able to identify shadows. (2Pl3)

Resources

A large space outside in the Sun; photocopiable page 26; card; coloured pencils; paints; paintbrushes; aprons; water; glue; sticky tape; small sticks (e.g. lollipop sticks or wooden skewers with the pointed ends cut off); torches and cells (batteries).

Starter

● Play the catching-shadows game again from the Starter activity on page 23.

● Ask the learners to make some different shadows (for example stretched, long, tiny). Ask them to look at and describe the shapes of the shadows they make.

● Identify shadows made from objects and buildings around them. Look at the shapes of these shadows and ask the learners to comment on what they see. Ask: *Are the shadow shapes the same or different from the object making them?*

Main activities

● Give out photocopiable page 26 and explain to the learners that they are going to make some shadow puppets. Go over the instructions on photocopiable page 26 in detail, step by step, and show them exactly what to do.

● Allow the learners to carry out the activity on photocopiable page 26 individually, or as a small group of learners who need extension activities.

● Bring the class back together to share what they have found out about making shadows. Explain that light travels in straight lines called 'rays'. It cannot go around corners or shine through solid things such as your body. On a sunny day you make a dark shape called a 'shadow' where the Sun cannot shine through you.

Plenary

● Explain that shadows are always the same shape as the object that makes them, but they can be longer or shorter, wider or narrower. Your shadow is the same shape as you and it will always copy the movements that you make.

● Ask some of the learners to demonstrate with their shadow puppets how to make the shadows bigger and smaller.

● Ask the learners who did the extension activity to perform their puppet show for the rest of the class.

Success criteria

Ask the learners:

● How is a shadow made? (Light is blocked.)

● What do you notice about the shape of shadows? (They are the same shape as the object blocking the light.)

● How can you make a bigger shadow?

● How can you make a shadow smaller?

● Can you escape from your shadow?

Ideas for differentiation

Support: Assist these learners to make their shadow puppets and direct them when investigating how to change the size of shadows.

Extension: Ask these learners to design and make puppets for a made-up or familiar story to perform to the rest of the class.

How to make shadows

You will need:

Card, coloured pencils (or paints, a paintbrush and an apron), scissors, a small stick, glue or sticky tape and a torch.

What to do

- Draw a shape of a person or an animal on the card.

- Colour or paint it and cut it out.

- Fix the wooden stick to the back of it.

- Take it outside so that the Sun is shining on it. If it is not sunny, use a torch in a dark room.

1. Look at the shadow your puppet makes.

2. Can you make it bigger?

3. Can you make it smaller?

4. Show your friends how you did it!

 Cambridge Primary: Ready to Go Lessons for Science Stage 2 © Hodder & Stoughton Ltd 2013

Shadow stories

Starter

- Ask the learners if any of them can make animal shapes with their hands to cast a shadow on a screen.
- Invite them to show the rest of the class. Let all the learners have a go. Some easy ones to make are a bird, by linking thumbs and waving the hands; a snake by simply wriggling the arm; a bird such as an ostrich by using the fingers as the top of the beak and the thumb as the lower part of the beak to make pecking movements.

Main activities

- Ask the learners to think with talk partners about what is making the shadow each time. Talk about making the shadows bigger and smaller, as in previous lessons.
- Ask the learners in pairs or small groups to make up a story or scene and act it out using their hands as puppets. Encourage them to use what they know about making shadows bigger and smaller to add interest to their stories. Have them perform to the rest of the class.
- Use a selection of different objects to cast shadows from behind a screen, for example a key, a plant, a guitar or mandolin (that is, a stringed instrument), a book, a shoe, a piece of fruit. Choose a selection of objects of which some are easily identifiable and others more difficult – a key has a distinct shape; a book can be viewed from different angles and therefore may be more difficult to guess. Try these out with the learners, asking them to guess the object each time.

- Repeat the above activity with the learners in pairs or small groups. Allow them to select their own objects to use.
- Give out photocopiable page 28 and explain to the learners that they have to look at the shadows and decide what could be making that shadow, then draw some shadows of their own making.

Plenary

- Invite a pair or group of learners to show their shadow story. Discuss how well they used their knowledge of making shadows.

Name: _____

What is making the shadow?

Look at these shadows and decide what is making them.
Write in the boxes. Make up two of your own.

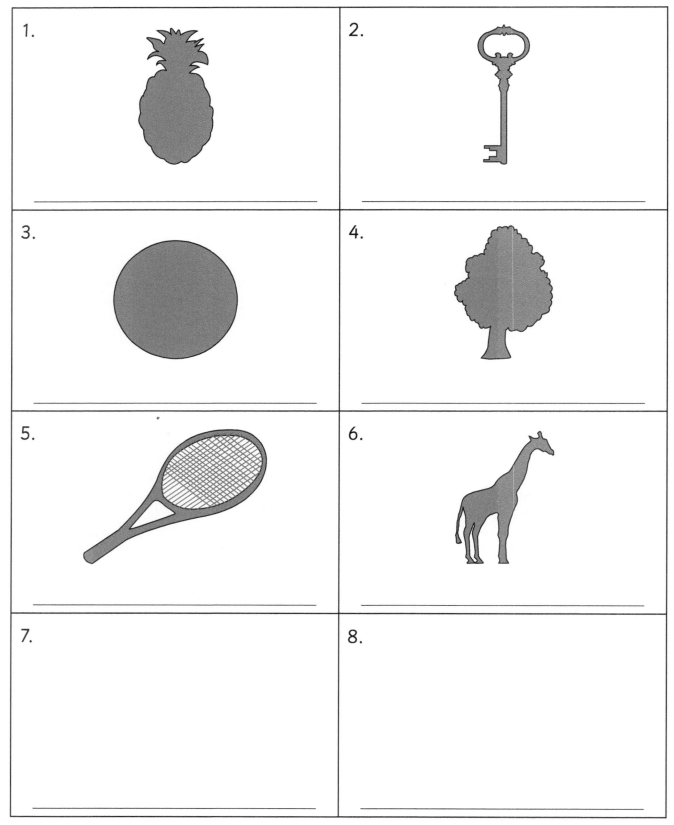

1.

2.

3.

4.

5.

6.

7.

8.

Investigating shadows 1

Learning objectives

- Ask questions and suggest ways to answer them. (2Ep4)
- Make and record observations. (2Eo3)
- Be able to identify shadows. (2Pl3)

Resources

Collection of objects from the previous lesson; photocopiable pages 28, 30, 31, 32 and 33; a selection of objects or materials that are either transparent or opaque, e.g. a block of wood or a wooden spoon, a block of plastic or a plastic toy, a piece of glass or a glass bottle or jar (under supervision), a block of metal or a key (from before).

Starter

- Look at and go over the learners' responses to photocopiable page 28 from the previous lesson. Discuss any misunderstandings.
- Ask some of the learners to show and talk about some of the shadows they made from objects that they chose themselves. Discuss if objects are easy or difficult to identify and why. (Easy to identify if the outline is clear and it is a distinctive object, more difficult if an item is a regular, solid shape or viewed from different angles.)

Main activities

- Explain that in this lesson the learners will find out what kinds of things make the best shadows.
- Ask the learners to tell you again what makes a shadow (light not being able to get through a solid object).
- Ask them to discuss with talk partners which materials or objects make good shadows.
- Guide them through thinking of an initial question, for example: 'What materials make the best shadows?' (Or give them this question to test.) Or encourage them to think of their own questions, such as: 'Is wood better at

making shadows than plastic?' This is the most difficult stage of scientific investigation, so the learners will need a lot of help and support to frame their thinking.

- Give out photocopiable pages 30 and 31 to all the learners except those who need support; give these learners photocopiable pages 32 and 33. Help all the learners to complete their photocopiable pages step by step. They should write down the question they are going to answer. Help them to make predictions. Look at their lists of equipment required and help them to decide how to record their results. At the end, encourage them to write down what they have found out and to see if they answer the initial question.
- Guide them to choose (or give them) two objects for making shadows. These must include a transparent and an opaque object – but do not use these words with the learners just yet. Suggest ways to make the test fair. Talk about how they might do the test – try each and compare the difference.

Plenary

- Invite the learners to show and tell their findings.
- Ask: *How did you make your test fair?*

Success criteria

Ask the learners:

- Are there any objects or materials that don't make shadows?
- Why is this?

Ideas for differentiation

Support: Ask these learners to complete photocopiable pages 32 and 33.

Extension: Ask these learners: *Which words describe objects that do not let any light through and objects that let light through?* (Opaque, transparent.)

Name: _____

Investigating shadows 1

My question

_____

_____ ?

I will need:

_____   _____

_____   _____

_____   _____

What I will do

(Draw or write in the box below.)

Name: _____

Investigating shadows 2

My results

(Draw or write what happened in the box below.)

What I found out

The _____ made the best shadow.

The _____ did not make a good shadow.

Name: _____

Which object makes the best shadow?

1. I think that the _____ will make the best shadow.

You will need:

Two objects given to you by your teacher, a torch and a screen.

What to do

teddy bear

torch

screen

• Make shadows using each object.

2. The objects I used were (draw or write):

1.	2.

Name: _____

Looking at shadows

1. What happened?

The shadows looked like this:

1.	2.

2. The shadows are different because:

_____

_____

3. The material that made the best shadow was _____

_____.

See-through or not?

Starter

- Play some shadow games – making shadows, chasing shadows, and so on, from the Starter activities on pages 25 and 27.
- Discuss making shadows change. Give the learners lots of opportunities to do and see this for themselves.
- Ask the learners in groups to make shadows of for example a table, a wheel, a letter of the alphabet. This is good for co-operation skills. Award points and perhaps prizes for good shadows.

Main activities

- Explain that in the last lesson the learners found out about which materials make good shadows.
- Ask them to remind you and each other which materials made the best shadows.
- Ask: *Why is this?* (They are solid and opaque and do not let any light through.)
- Look at the clouds, or a picture of a cloudy day.
- Ask the learners to discuss with talk partners what difference clouds make to sunshine on a cloudy day (it is dull because the clouds stop some of the sunlight breaking through; it can also be colder when this happens).

- Explain that in this lesson they will predict which things let light through, which let some light through and which do not let light through. Remind them that a prediction is a guess.
- Give out photocopiable page 35 and explain to the learners that first of all they are going to predict if things will let a lot, some or no light pass through them. Then they will be able to try it out for themselves to see if they were right.
- Allow them to complete the table at the top of photocopiable page 35 to show their predictions.
- Check this before allowing them to test the objects and materials.

Plenary

- Go through the learners' responses. Introduce the words 'transparent' and 'opaque' but do not expect all the learners to remember these words. The learners who need extension may also be introduced to the word 'translucent', but again, do not expect them to remember this word. Learners just like to use some real scientific words.

Name: _____

See-through or not?

1. Complete the table with your predictions. Tick (✓) to show what you think.

Material or object	Lets light through	Lets some light through	Does not let light through
glass			
tissue paper			
sunglasses			
wood			

You will need:

A torch and a screen.

What to do

- Try to make shadows using the things in the table.

- Tick (✓) to show what happened.

Material or object	Lets light through	Lets some light through	Does not let light through
glass			
tissue paper			
sunglasses			
wood			

Unit assessment

- What is the main light source for us on Earth?
- Is a torch a natural or an artificial source of light?
- Name another natural light source.
- Name a different artificial source of light.

- What makes a shadow?
- What happens to the size of a shadow when you move the object nearer to the light source?

Summative assessment activities

Observe the learners while they participate in these activities. You will quickly be able to identify those who appear to be confident and those who may need additional support.

Light sources

This activity assesses the learners' knowledge of different light sources, and their classification as natural or artificial.

You will need:

A selection of different light sources – as used in early lessons in this unit, or a selection of pictures of different light sources.

What to do

- Working independently with you, ask each learner to name the light source that you point to.
- Ask each learner to group the sources into natural and artificial (man-made).
- Record the learners' responses on a class checklist and use the information when writing reports or feeding back to the learner and their parents.

Making shadows

This activity assesses the learners' understanding of how to make and alter shadows.

You will need:

The creative / language area of the classroom needs to have a shadow-puppet theatre set up permanently for the duration of this unit; if this is not usual procedure, set one up for a few days or a week to provide opportunities to observe the learners interacting on this activity.

What to do

- Prepare some cards with questions on for display in the area of the puppet theatre, for example: 'What is the biggest shadow you can make?', 'Can you make the puppet run away?', 'Can you show the puppet coming closer?'
- Take time to observe the learners at play. Interact with them and ask some of the questions at the top of this page to prompt their thinking.
- Record the learners' responses on a class checklist and use the information when writing reports or feeding back to the learner and their parents.

Distribute photocopiable page 37. The learners should work independently, or with the usual adult support they receive in class.

Name: _____

Day and night

Use words from this list to help you complete the sentences below.

dark   day   Earth   light   night   Sun   sunrise   sunset

1. We wake up when it is _ _ _ _ _.

2. If there is no light we say it is _ _ _ _.

3. The _ _ _ gives us daylight on _ _ _ _ _.

4. The Earth spins on its axis: this gives us _ _ _ and

    _ _ _ _ _.

5. At the start of the day when the Sun appears it is called

    _ _ _ _ _ _ _.

6. At the end of the day when the Sun disappears, it is

    called _ _ _ _ _ _.

Looking at a torch

Learning objectives

● Collect evidence by making observations when trying to answer a science question. (2Ep1)
● Use first-hand experience. (2Ep2)
● Recognise the components of simple circuits involving cells (batteries). (2Pm1)

Resources

Torch(es) – with the cells (batteries) inserted the wrong way round; photocopiable pages 39 and 40; brochures or magazines containing pictures of electrical items; paper; scissors; glue.

Starter

• Demonstrate with a single torch or allow the learners to work in pairs to make a torch light up. Allow them to change the cells (batteries) or lamps if they think this might make a difference.
• Ask the learners to show and tell the rest of the class what they had to do to make the torch light up (put the cells [batteries] in the correct way). Ensure that all the torches are now working properly.

Main activities

• Allow the learners in pairs or small groups to disassemble the torch (take it to pieces) to find out how it works. Ask: *Can you put it back together at the end so that it is still working?* Invite different pairs or groups to show you how they do this.
• Guide the learners systematically through the flow of electricity in the torch circuit, for example ask:
  • *Where does the electricity come from?* (The cell [battery].)
  • *What do you have to do to make the torch light up?* (Push a button or slide a switch.)
  • *How do you think the switch is joined to the cell (battery)?* (By wires.)
  • *What happens when you switch (turn) the torch off?* (It stops the electricity from the cell [battery] getting to the lamp to make it light up, and so the light goes off.)

• Explain that the things that make the torch light up are the same things that make anything light up that runs off cell (battery) power. Think of things that do this, for example car headlights, festival lights, a laptop screen.
• Ask the learners to tell you the names of any of the parts of the torch, for example cell (battery), lamp, switch, reflector, wires, case, handle.
• Give out photocopiable page 39 and ask the learners to draw and label a picture of a torch; use photocopiable page 40 for the learners who need support (this has a picture of a torch on it and words to help them label the parts correctly).

Plenary

• Go through the correct responses to photocopiable pages 39 and 40 with the learners.
• Talk about the different parts of the torch.
• Identify that the lamp lights up, the switch turns it on and off, and the cell (battery) supplies the electricity. Electricity moves through wires to make the lamp light up.

Success criteria

Ask the learners:

● Where is the power source for the electricity in the torch? (The cell [battery].)
● What does the switch do? (Turns the light on or off.)
● What takes the electricity from the cells (batteries) to the lamp? (Wires.)

Ideas for differentiation

Support: Ask these learners to complete photocopiable page 40.

Extension: Ask these learners to make a collage of different pictures of items that use electricity.

Name: _____

Looking at a torch

1. Draw and label a picture of a torch. Name as many parts as you can.

2. Write a list of things that you need to make something light up.

   _____

   _____

   _____

   _____

Name: _____

Looking at a torch

1. Look at this picture of a torch.

2. Use these words to help you label as many parts as you can.

lamp	handle	light	switch	wire

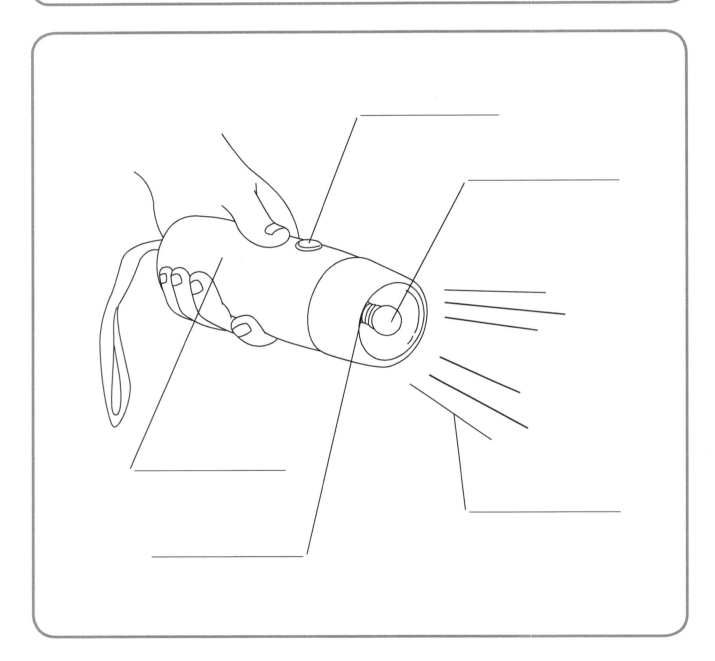

3. What makes the torch light up?

_____

_____

Cambridge Primary: Ready to Go Lessons for Science Stage 2 © Hodder & Stoughton Ltd 2013

Making circuits

Learning objectives

- Predict what will happen before deciding what to do. (2Ep5)
- Make and record observations. (2Eo3)
- Recognise the components of simple circuits involving cells (batteries). (2Pm1)

Resources

Torches; photocopiable pages 42 and 43; simple electrical circuit components – cells (batteries) and cell (battery) holders; wires; lamps; lamp-holders.

Starter

- Recap from the previous lesson what components were needed in a torch to make it light up – cells (batteries), arranged in the right order; a lamp; some wires.
- Give out some torches and ask some of the learners to identify these parts on the torch.
- Give out some torches that work and some that don't work. Ask the learners to make them all light up.

Main activities

- Show the learners the separate circuitry components – lamps, lamp-holders, cells (batteries), cell (battery) holders and wires. Pass them around and let the learners handle the equipment.
- Demonstrate how to put lamps in the lamp holders and cells (batteries) in the cell (battery) holders. Allow the learners to practise doing this for themselves. (Sometimes such equipment is fiddly or the learners think that they might break it if they push too hard.)
- Ask the learners in pairs or small groups to use only the given equipment – one cell (battery), one cell (battery) holder, one lamp, one lamp-holder and two wires. Give out photocopiable pages 42 and 43 and explain that they have

two jobs to do. They need to make a circuit that **will not** make the lamp light up and a circuit that **will** make the lamp light up. Introduce the word 'circuit' as the journey of the electricity from the cell (battery) to the lamp and back to the cell (battery) again. Describe it as a circular journey, to convey the notion of electricity flowing / travelling around.

- Carry out the activities described on photocopiable pages 42 and 43. Move round the classroom as the learners carry out these activities. Guide, support and suggest to them ideas to help them make the lamp light up or not. Some of them will easily be able to make the lamp light up.

Plenary

- Invite different groups of learners to show the rest of the class the ways in which they made the lamp light up.
- Repeat this, showing how not to make the lamp light up.
- Recap that the circuit needs to be complete for the lamp to light up.

Success criteria

Ask the learners:

- How did you arrange the equipment to make the lamp light up?
- How did you change the circuit to prevent the lamp from lighting up?
- What do you need to make the lamp light up? (A complete circuit.)

Ideas for differentiation

Support: Either work in a small group, supporting these learners, or allow them to work in mixed-ability groups.

Extension: Challenge these learners to make the lamp go on and off.

Name: _____

How to make circuits 1

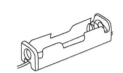

Work in a pair or small group.

You will need:

a cell (battery)

a cell (battery) holder

two wires

a lamp

a lamp holder

What to do

- Make and draw two circuits. Draw one in the box below and the other on the 'How to make circuits 2' page.

How to make the lamp light up

Cambridge Primary: Ready to Go Lessons for Science Stage 2 © Hodder & Stoughton Ltd 2013

Name: _____

How to make circuits 2

When the lamp does not light up

What three things are needed to make the lamp light up?

1. _____

2. _____

3. _____

Looking at switches

● Make comparisons. (2Eo6)
● Identify simple patterns and associations. (2Eo7)
● Know how a switch can be used to break a circuit. (2Pm2)

Flipchart and markers or interactive whiteboard; torches; circuitry equipment – lamps and lamp-holders; cells (batteries) and cell (battery) holders; wires; switches; photocopiable page 45.

Starter

• Ask the learners who managed to carry out the extension activity in the previous lesson to show and tell the rest of the class what they did and how they made the lamp go on and off.

• Ask the learners to discuss with talk partners if there is a better way to control the light.

• Discuss and list the learners' responses. Explain that they will have the opportunity to try out some of their ideas later in the lesson.

Main activities

• Give out the torches again and ask the learners to switch / turn the torch on and off. Discuss the job that the switch does; that is, it controls the light, or switches / turns it on and off.

• Explain that there are switches available that can be incorporated into their circuits. Show and demonstrate the different kinds of switches available for the learners to use.

• Give them time to look at and handle the switches. Talk about how they could be used in a circuit.

• Explain that they are going to have the opportunity to make a circuit and include a switch in it.

• Remind them from the previous lesson that the circuit needs to be complete for the electricity to flow and make the lamp light up.

• Ask them in pairs or small groups to re-make a circuit using just a cell (battery) and cell (battery) holder, a lamp and a lamp-holder, and some wires. Check that all their circuits are correct and working before continuing.

• Give out photocopiable page 45 for the learners to complete. Explain that they need to make a circuit including a switch and be able to draw what they have made.

• Circulate as the learners carry out the activity. Provide help, advice and assistance to enable them to be successful in the task. Discuss any errors or misconceptions as they arise.

Plenary

• Ensure that all the groups of learners have successfully made or seen a circuit controlled by a switch.

• Ask some of the groups to show their circuits to the rest of the class.

• Discuss how a switch works – it breaks the circuit when it is open and so stops the flow of electricity.

Ask the learners:

● What difference does including a switch in a circuit make?
● How did you make your switch work? (Some switches slide; others press, push or twist.)
● What does the switch do when it is open? (Stops the lamp from lighting up.)
● Why does the lamp light up when the switch is closed? (The circuit is complete.)

Support: Assist these learners by working with them in a small group. Alternatively, organise them to work in mixed-ability groups.

Extension: Challenge these learners to make a mask and make the eyes light up.

Name: _____

Circuits with switches

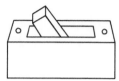

A switch helps to control a circuit.

A switch can be used to turn a lamp on and off.

You will need:

a cell (battery)

a lamp

a switch

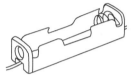

a cell (battery) holder

a lamp holder

wires

What to do

- Make a circuit, including a switch.
- Draw and label your circuit.

Is the lamp lit in your diagram? yes / no

Finding switches

- Collect evidence by making observations when trying to answer a science question. (2Ep1)
- Use first-hand experience. (2Ep2)
- Know how a switch can be used to break a circuit. (2Pm2)

Resources

Circuitry equipment; different toys and everyday items that use switches, e.g. a light, a lamp, a torch, a computer, a calculator, a mobile phone, a digital camera; photocopiable page 47.

Starter

- Have a ready-made circuit prepared, including a switch and a lamp. Set the circuit up so that there is a problem with it, for example the cells (batteries) are inserted the wrong way round, or one of the wires is not connected, and so on.
- Ask the learners to predict what will happen when they use the switch.
- Invite a learner to try the switch. Noting the learners' reactions, ask them to suggest how to make the circuit work to make the lamp light up.
- Alternatively, have the same circuit set up for each small group. Make the problem with the circuit different on each table. Ask the learners in pairs or small groups to identify the problem with the circuit and restore it to working order. This is a good activity to reinforce previous learning about making circuits.
- In turn, ask the learners to show how they made the circuits work properly.

Main activities

- Explain that all the components in a circuit have to be in good working order. The switch then controls the circuit. Remind the learners that a switch works by making or breaking the circuit.

- Ask the learners to think with talk partners about as many different things as they can that have switches on them.
- Discuss the types of things they have already seen and used today that have switches.
- Explain to the learners that they are going to find and identify different kinds of switches. Give out photocopiable page 47 and tell them that they are going to find, draw or identify different kinds of switches.

Plenary

- Ask the learners to go around the room and show some examples of switches that they have found.
- Discuss how each of these switches works – is it a pull, a push, a slide or a press?
- Go over the responses to photocopiable page 47, noting their answers.

Success criteria

Ask the learners:

- What is the job of a switch in a circuit? (It turns the light on or off.)
- What different ways have you found to work a switch? (Push, pull, slide, press.)
- Can you name something that works using a push / pull / slide / press switch?
- If the switch is working, but the circuit isn't – what else might be wrong with the circuit?

Ideas for differentiation

Support: Assist these learners by working with them as a small group. Help them to choose the correct descriptive word for how the switches work.

Extension: Give these learners a buzzer instead of a lamp. Ask: *Can you make a circuit, including a switch, that makes the buzzer work?*

Name: _____

Different switches

1. Find three different things around the room that have a switch.

   a) _____

   b) _____

   c) _____

2. Use these words to help describe how the switches below work.

| pull | push | press | slide |

Object	How the switch works
	_____ _____ _____
	_____ _____ _____
	_____ _____ _____
	_____ _____ _____

Different types of cells (batteries)

Learning objectives

- Collect evidence by making observations when trying to answer a science question. (2Ep1)
- Use first-hand experience. (2Ep2)
- Recognise the components of simple circuits involving cells (batteries). (2Pm1)

Resources

A variety of different cells (batteries), e.g. cells, cylindrical, 9V, mobile phone, digital camera; items that use the different types of cells (batteries), e.g. calculator, torch, mobile phone, digital camera; photocopiable page 49.

Starter

- Tell the learners that just as switches can be different (as they found out in the previous lesson), so there are many types of cells (batteries).
- Think back to the items on photocopiable page 47. Ask the learners if any of them know anything about what type of cells (batteries) these objects use.
- Show some of the different types of cells (batteries) and ask the learners to predict which objects need that particular kind of cell (battery).
- Discuss their responses – find out how they know (for example they might have one like it at home).
- Tell them that they will have the opportunity in this lesson to find out about different types of cells (batteries).

Main activities

- Discuss the names of the different kinds of cells (batteries), for example those listed in the resource section.
- Give the learners in pairs or small groups an object or group of objects to find out what kind of cells (batteries) they use. Alternatively, set up a carousel of activities where the learners visit different tables in turn, each containing a particular object. As the lesson progresses, have them visit each table in turn until they have examined all the objects.

- Emphasise that they just need to look at them – not touch or remove them, especially mobile phone cells (batteries) – and identify which type they are.
- Give out photocopiable page 49 and explain to the learners that they have to identify different types of cells (batteries) and then to match each one to an item that uses that particular type of cell (battery).

Plenary

- Go over the learners' responses to photocopiable page 49. Discuss their answers and show them the actual cells (batteries).
- Discuss any misunderstandings as they arise.

Success criteria

Ask the learners:

- What is the power source for circuits that do not use mains electricity?
- Tell me the names of some different types of cells (batteries).
- What kind of cell (battery) is this?
- What kind of equipment would use a cell (battery) like this one?
- What kinds of equipment have special cells (batteries)?

Ideas for differentiation

Support: Assist these learners in a small group when looking at equipment and the cells (batteries) inside.

Extension: Ask these learners: *For each type of cell (battery) shown on photocopiable page 49, can you name another item that uses that type of cell (battery)?*

Name: _____

Different cells (batteries)

Here are some different types of cells (batteries).

Draw a line from each object to show the type of cell (battery) that it uses.

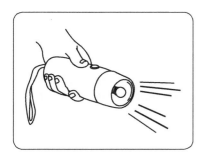

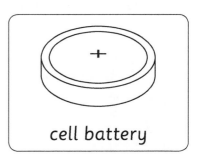

cell battery

10:24

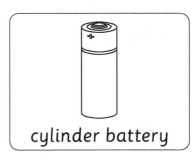

cylinder battery

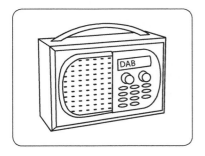

DAB

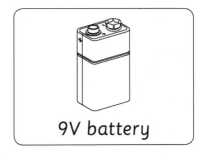

9V battery

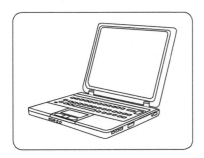

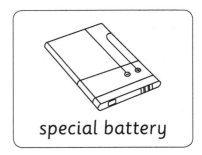

special battery

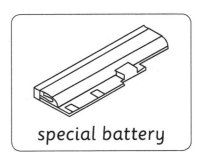

special battery

Cells (batteries) or mains electricity?

Learning objectives

- Use simple information sources. (2Ep3)

Resources

Photocopiable pages 51 and 52 (one for each pair or group); flipchart and markers or whiteboard; some (or all) of the items shown on the playing cards.

Starter

- Copy photocopiable page 51 onto card and make sets of cards. Laminate them if possible.
- Sort the learners into pairs or small groups. Give out the instructions on photocopiable page 52 for them to follow and play the game. Tell them that you expect them to read and follow the instructions to play the game.
- Allow them to play the game.

Main activities

- Go through the learners' responses to the Starter activity game.
- List the items that use cells (batteries) or not for the learners to refer to throughout the lesson.
- Discuss until agreement is reached. Please note: some of these items can run from cell (battery) power or mains electricity – for example a torch, hand-held game, laptop. This should generate some good discussion.
- Talk about each list in turn, starting with the 'cell (battery)' list. Discuss who has these things at home, perhaps how long the cells (batteries) last or how long they take to re-charge.
- Perhaps do a quick 'hands up' survey for the learners to indicate by a show of hands which of these items they have at home.
- Then go on to discuss the 'not' list. Discuss which of these items the learners are familiar with. Talk about the difference between how these items work and the items in the 'cell (battery)' list.

- Introduce the words 'mains electricity'. Define this as electricity that comes from a plug inserted into an electrical socket and switched on.

Plenary

- Play the Starter activity game again as a class – hold up the cards for the learners to respond with the answers 'cell (battery)' or 'mains'.

Success criteria

Ask the learners:

- How many things did you find in the game that use cell (battery) power?
- What do we call the power source if something doesn't use a cell (battery)?
- Tell me something that uses cell (battery) power.
- What items at home, in the kitchen, use mains electricity?
- What do you need for mains electricity to work? (Sockets and a mains power supply to the home.)
- Are most electrical items at home cell (battery) powered or mains electricity?

Ideas for differentiation

Support: Play the Starter game in a small group with these learners. Talk to them about the things that can run on both cell (battery) and mains electricity.

Extension: Challenge these learners to find and label three things around the classroom that use cell (battery) power and three things that use mains electricity.

Cell (battery) or mains?

Name: _____

Cell (battery) or not?

You will need:

A set of playing cards, a partner or small group.

What to do

- Place all the cards in a pile face down on the table.
- Take it in turns to pick a card.
- Look at the picture. Does it use a cell (battery) or not?
- Decide and put it in one of the boxes below.
- Keep taking turns until all the cards have been used up.

Cell (battery)	Not

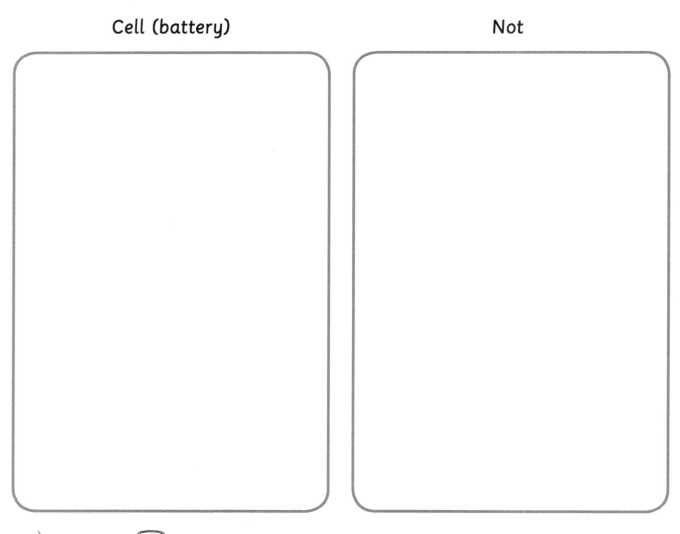

Cambridge Primary: Ready to Go Lessons for Science Stage 2 © Hodder & Stoughton Ltd 2013

What uses electricity at home?

Starter

• Label opposite corners of the classroom 'cell (battery)' and 'mains'. Invite the learners to choose a card from the pile and go to the correct corner. In each corner, the learners should display their pictures from the cards.

• Review the pictures with the learners and discuss until agreement is reached. Invite the learners with the cards to justify why they placed the card in the 'cell (battery)' or 'mains' corner.

• Ask the learners to remind you again what mains electricity is.

• With talk partners, ask the learners to think of as many things as they can that run from mains electricity at home.

• Listen and respond to some of their answers; discuss further any responses that are unusual.

Main activities

• Explain that in this lesson the learners are going to pretend to take a walk around their home, or a room at home, and to remember as many things that work on mains electricity as they can.

• Ask the learners to close their eyes. In words, take them for a walk through their house, or a room, for example: *Imagine you are in your bedroom ... what can you see there that uses mains electricity? What is your favourite thing?*

• Ask the learners to open their eyes and to share with the rest of the class some of the things that they have remembered.

• Write a list of useful words of things that the learners mention – this will help them to complete the photocopiable pages. Display the list in a prominent place for them to refer to.

• Give out photocopiable page 54 to all except the learners who need extension. For the learners who are completing photocopiable page 54, either let them choose a room in their house to draw, or choose for them. Explain that they need to draw and label a room in their house with electrical items in it.

• Give photocopiable page 55 to the learners who need extension. Explain that they need to think about rooms all over their house to draw and label.

Plenary

• Invite the learners to show and talk about their work with the rest of the class.

• Think about which rooms have a lot of electrical items and which do not.

• Display a set of photocopiable page 54 in a similar format to that shown on photocopiable page 55 to make up an imaginary home from a combination of four different learners' work. Display this prominently.

Name: _____

Electricity at home

Many things in our homes use electricity.

1. Choose a room in your home.

2. Draw and label a picture that shows as many things as you can that use electricity.

3. Use the class word list to help you.

My _____

How many things did you find? _____

Cambridge Primary: Ready to Go Lessons for Science Stage 2 © Hodder & Stoughton Ltd 2013

Name: _____

Electricity at home

1. In this house, draw and label as many things as you can that use electricity.

2. Use the class word list to help you.

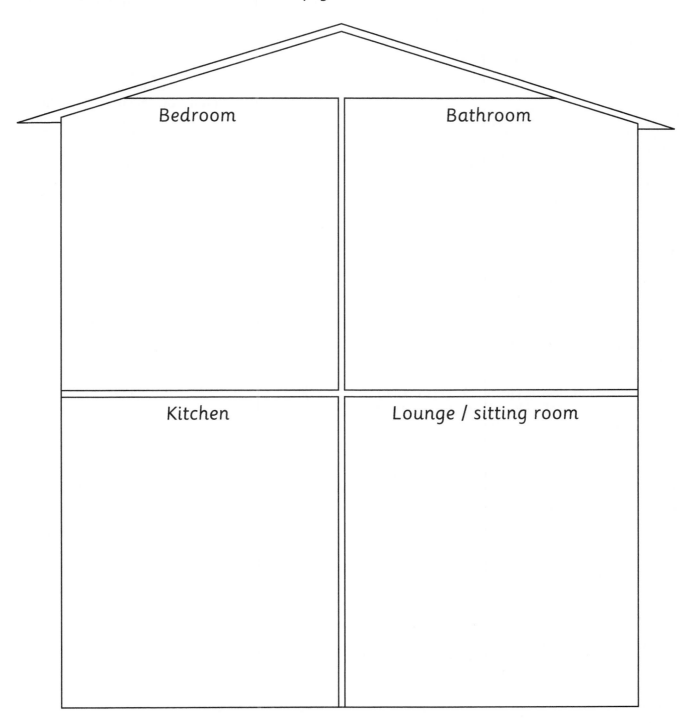

| Bedroom | Bathroom |
| Kitchen | Lounge / sitting room |

3. Which room uses the most electricity? _____

Circuits for light, sound and heat

- Collect evidence by making observations when trying to answer a science question. (2Ep1)
- Make and record observations. (2Eo3)
- Recognise the components of simple circuits involving cells (batteries). (2Pm1)

Internet access; circuitry equipment, including buzzers and motors; a lamp; a stereo or radio; a digital camera or TV; an electric fan; a toaster; photocopiable pages 57, 58 and 59.

Starter

- If you have internet access, play 'Electro-matic factory'. Go to www.scibermonkey.org: ages 5–7 → Energy → Electricity → Electro-matic factory. Choose the medium activity. This asks the learners to choose which equipment uses electricity to make light (a lamp), sound (a stereo or radio), pictures (a digital camera or TV), movement (an electric fan) and heat (a toaster). It also introduces the idea of electrical safety in that a bonus activity asks the learners to identify an unsafe thing (an overloaded socket). If you do not have internet access, a similar activity could be used in class, using similar questions and familiar everyday electrical items.
- Re-cap with the learners that the electricity in circuits can be used to make light (lamps), sound (buzzers), pictures (TVs and digital cameras), things move (motors) and to produce heat (toasters).

Main activities

- Explain that in this lesson the learners will have the opportunity to consider what electrical components can be used to generate light, sound or heat.
- Remind them that they have already made circuits to make a lamp light up and circuits with switches. The learners who need extension activities may also have used a buzzer.

- In pairs or small groups, ask the learners to re-make a circuit, including a switch that makes a lamp light up, and show the rest of the class.
- Ask the more-able learners to show a circuit including a buzzer, if possible. If they have not tried this already, give them the opportunity to have a go now. Then ask them to show the rest of the class how they did it.
- Give out photocopiable page 57 to the learners who need support, photocopiable page 58 to the main group and photocopiable page 59 to those learners who need extension. Explain that they need to think about what kind of circuit makes a particular effect. Some learners need to think about what is in the circuit for these effects to happen.

Plenary

- Go through the learners' responses to photocopiable pages 57, 58 and 59.
- Invite the learners from each group to discuss their responses.
- Remind them that electricity can be used to make useful effects – for example light, sound, pictures, movement and heat.
- Ask the learners who completed photocopiable page 59 to identify which components produce such effects.

Ask the learners:

- How can you tell if a lamp is working?
- Tell me something that uses electricity to make sound.
- How is electricity useful in a TV?
- Describe how electricity can be used to create movement.

Support: Give these learners photocopiable page 57 to complete.

Extension: Give these learners photocopiable page 59 to complete.

Name: _____

Using electricity

Circuits use electricity to do lots of useful things for us.

Circuits can be used to make:

light	pictures	sound

1. Look at these pictures of equipment.

2. Use one of the words from the box above to show how it uses electricity.

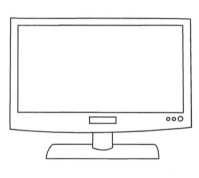

_____   _____   _____

3. Draw three things of your own that use electricity.

4. Label them to show how they use electricity.

_____   _____   _____

Using electricity

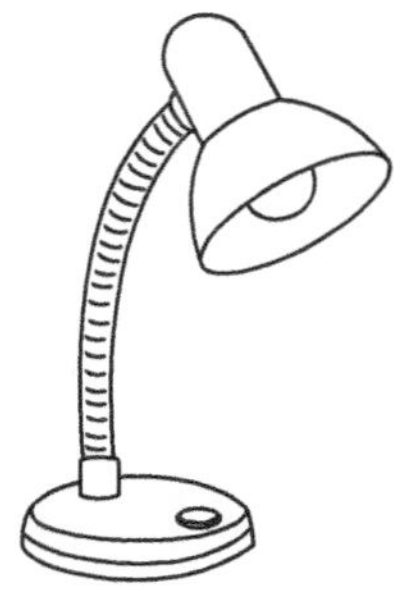

Circuits use electricity to do lots of useful things for us.

1. Draw a picture of something that uses electricity to make these things.

heat

light

movement

pictures

sound

Cambridge Primary: Ready to Go Lessons for Science Stage 2 © Hodder & Stoughton Ltd 2013

Name: _____

Using electricity

Circuits use electricity to do lots of useful things for us.

1.   (Circle) the items that use mains electricity.

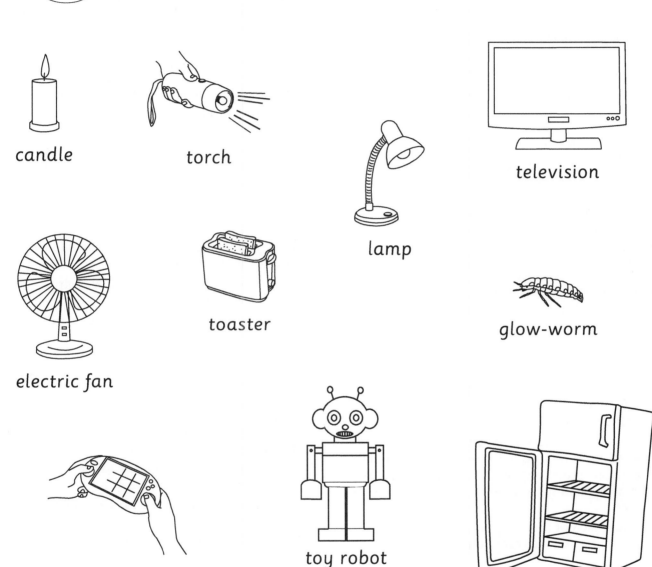

candle   torch   television

lamp

electric fan   toaster   glow-worm

hand-held   toy robot   refrigerator
electronic game

2.   Write the names of these components.

 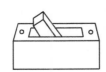 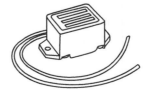 

_____   _____   _____   _____

Electrical safety

- Talk about risks and how to avoid danger. (2Eo2)
- Identify simple patterns and associations. (2Eo7)
- Know how a switch can be used to break a circuit. (2Pm2)

Pictures (internet or books) of possible dangerous situations involving electricity – for example damaged plug, overloaded sockets, baby left unattended near electrical equipment, knife being used to retrieve toast from a toaster, someone unplugging a device by pulling the cable; poster paper and coloured pencils or computer access to a design package; photocopiable pages 61 and 62.

Starter

- Ask the learners to think with talk partners about ways in which we keep safe from electricity at home.
- Discuss the learners' responses (for example use socket protectors, do not leave babies alone near electrical equipment, don't have trailing cables, don't overload sockets, don't use electricity near water).
- Discuss each of their responses, asking how these things keep us safe from electricity.

Main activities

- Use pictures to discuss any safety situations that the learners have not identified during the Starter discussion. It is likely that these will need longer discussion, as these are things that did not readily come to mind for them.
- Alternatively, set up a room scene in the classroom and ask the learners to identify any electrical hazards that they can see.

- Explain that mains electricity is much more powerful than electricity that comes from a cell (battery). This is why we use cells (batteries) to power circuits we make in the classroom. Mains electricity can cause serious burns and shocks and even kill people. It is important to be safe around electrical equipment, and this is what they are going to remind themselves of and perhaps find out more about in this lesson.
- Ask them to design and make a poster warning others how to be safe with electricity.
- Give out photocopiable page 61 and explain to the learners that they have to look at the picture and find the hazards.

Plenary

- Ask the learners to identify the hazards they found on photocopiable page 61.
- Invite some of the learners to show their electrical-safety posters.
- Discuss good places to display these around school.

Ask the learners:

- How should you unplug something? (Use the plug, not the cable of the appliance.)
- Which electrical items are safe to use near water? (None.)
- Is it safe to put a knife in a toaster if the toast gets stuck?
- What is the difference between cell (battery) power and mains electricity?
- Tell me something that is unsafe as far as electricity is concerned.

Support: Assist these learners in identifying the hazards on photocopiable page 61. Give support with lettering and spelling (if any) on the posters.

Extension: Give these learners the wordsearch on photocopiable page 62 after they have finished photocopiable page 61.

Name: _____

Using electricity safely

Look closely at the picture and circle all the hazards that you can find.

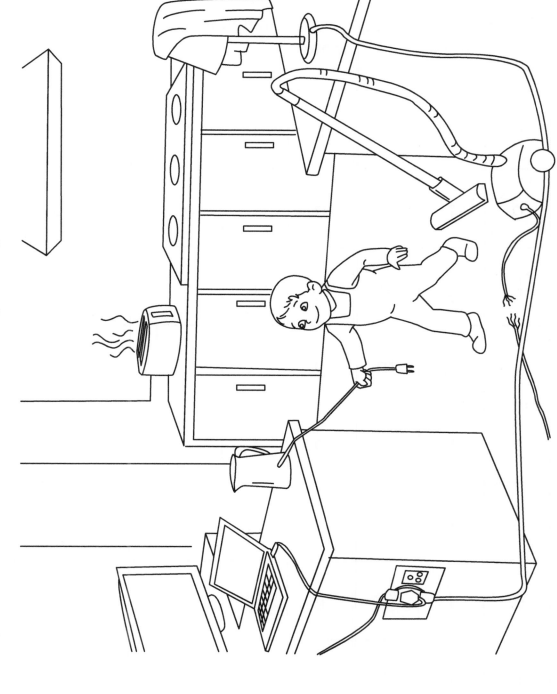

Name: _____

Electricity wordsearch

Find the words below. They read across or down.

| buzzer | circuit | component | electric | lamp |
| light | motor | switch | wires | |

a	s	b	c	i	r	c	u	i	t
d	f	m	o	t	o	r	o	e	p
k	y	e	m	t	r	b	s	l	b
l	a	m	p	a	l	u	n	e	z
f	r	l	o	t	i	z	i	c	a
s	w	y	n	a	g	z	o	t	p
w	i	r	e	s	h	e	f	r	g
h	l	i	n	k	t	r	l	i	m
s	w	i	t	c	h	n	p	c	o

Cambridge Primary: Ready to Go Lessons for Science Stage 2 © Hodder & Stoughton Ltd 2013

Closing the circuit

Learning objectives

- Identify simple patterns and associations. (2Eo7)
- Recognise the components of simple circuits involving cells (batteries). (2Pm1)
- Know how a switch can be used to break a circuit. (2Pm2)

Resources

Internet access; circuitry equipment; a selection of other materials, e.g. paperclips, stones, plastic or wooden rulers, coins, drinking straws, a leaf; photocopiable pages 64 and 65.

Starter

- Ask the learners in pairs or small groups to make a circuit with a switch to make a lamp light up.
- Check that all the circuits have been constructed correctly and are in good working order.
- Remind the learners that the switch helps to control the flow of electricity in the circuit.
- Remind them of the terms 'making the circuit' and 'breaking the circuit'.
- Ask the learners what they have to do to make the circuit using the switch (close the switch).
- Ask the learners to remove the switch from their circuit, but to leave a gap where the switch has been.
- Ask: *Will the lamp light now?* (No.)
- Ask the learners what they have to do to make the circuit work without the switch (join the circuit / close the gap).

Main activities

- Explain that in this lesson the learners are going to investigate if they can replace the switch with other items and see the effect that this has on the circuit.
- Show them some of the items available to use, as listed in the resources section.

- Ensure that all the learners know that whatever they use, they must make sure that it is touching both wires where the switch has been removed.
- Explain that they will have the opportunity to try different things in the gap, and then they will be expected to report back to the rest of the class.
- Either pre-select – for example three different objects for each group to test – or allow them all to try the same three objects.
- For the learners who need support, give them photocopiable page 64. Give all the other learners photocopiable page 65. Tell these learners that they have to choose what to put in the table.

Plenary

- Go to www.scibermonkey.org: age 5–7 → Energy → Electricity → Using Electricity. This activity allows the learners to predict and then try what happens when different materials are used in a gap in a circuit. It also explores what effect changing the cell (battery) around in the circuit makes (none). If internet access is not available, this can be done using actual circuitry equipment in class, using the materials tried in the lesson and perhaps a few that have not been tried.

Success criteria

Ask the learners:

- Which material did you try first that worked?
- Name a material that did not make the circuit.
- What do you notice about the kinds of materials that do complete the circuit?

Ideas for differentiation

Support: Give these learners photocopiable page 64 to complete.

Extension: Ask these learners to make a generalisation about which type of materials are good at making the circuit (metals).

Name: _____

Completing the circuit

You made a circuit like this one and removed the switch.

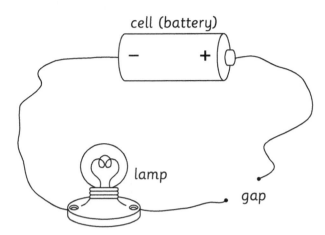

cell (battery)

− +

lamp

gap

Complete the table to show which things you used instead of the switch.

Material or object	Did the lamp light up? Tick (✓) if it did

Cambridge Primary: Ready to Go Lessons for Science Stage 2 © Hodder & Stoughton Ltd 2013

Name: _____

Completing the circuit

1. Draw the circuit that you made, with the gap left where you removed the switch.

2. Complete the table to show what you found out.

Material or object	Did the lamp light up? Tick (✓) if it did

Unit assessment

- What is the power source for an electric circuit?
- What do we call two or more cells (batteries) in a circuit?
- How can we tell if a circuit is working?
- What can we use to control a circuit – for example to make a lamp light up or turn off?

- What equipment would you need to make a lamp light up?
- What could you use instead of a switch if there was a break in the circuit?

Summative assessment activities

Observe the learners while they participate in these activities. You will quickly be able to identify those who appear to be confident and those who may need additional support.

Making circuits

This activity assesses the learners' ability to construct working circuits.

You will need:

Basic circuitry equipment; digital camera (if available).

What to do

- Set up a permanent activity table in the classroom during this unit of work, if possible. If not, set up an activity table or make the circuitry equipment readily accessible for the learners to choose during this activity. Invite the learners to play and make circuits and to take photographs of working circuits they make.
- When possible, observe the learners during this activity to gain a sense of their use of appropriate scientific language.
- Use the photographs as evidence that they can make working circuits. Use this in learning journals or as a record of achievement, if necessary.

Making switches

This activity allows the learners to design and make their own switch to incorporate into a circuit.

You will need:

Basic circuitry equipment; aluminium foil; paperclips; card; drawing pins (thumb tacks); digital camera (if available).

What to do

- Working individually or in pairs, ask the learners first to make a working circuit that makes a lamp light up that is controlled by a switch.
- Record on a class checklist those who can do this.
- Ask them to remove the switch and either make another switch from the other materials available or simply insert something into the circuit that will make the lamp light again.
- All the learners should be able to complete the circuit.
- This switch-making activity differentiates, allowing the learners who need extension activities to investigate making switches. Take photographs as evidence of any working switches made.

Distribute photocopiable page 67. The learners should work independently, or with the usual adult support they receive in class.

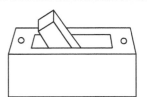

Name: _____

Making circuits

1. Draw and label a circuit containing a lamp, wires and a cell (battery).

2. Draw and label a working circuit that uses a switch.

Changing shape

Learning objectives

- Use first-hand experience. (2Ep2)
- Make and record observations. (2Eo3)
- Know how the shapes of some materials can be changed by squashing, bending, twisting and / or stretching. (2Cc1)

Resources

A large indoor or outdoor space; modelling clay, dough or salt dough; rubber bands; plastic drinking straws; pieces of paper; bath sponges; woolly scarves; photocopiable page 69.

Starter

- Go into a suitable large space, for example the playground or PE hall. Introduce the learners to the important vocabulary associated with this chemistry topic, such as 'squash', 'bend', 'twist' and 'stretch'. Ask the learners to run around and, on your command, stop and make a particular kind of shape, such as:
  - *Squash* yourself into the smallest possible shape. Squash yourself flat against the wall or floor.
  - *Bend* and touch your toes. Bend like a tree or grass blowing in the wind.
  - *Twist* and turn slowly, three times on the spot. Stand still and twist the top half of your body to face the ...
  - *Stretch* and make the tallest shape that you can. Stretch your body and make the widest shape that you can.

  This will familiarise the learners with language that they will need to remember.

Main activities

- To make salt dough, mix one cup of salt, two cups of flour and one cup of lukewarm water and knead. It can be stored in an airtight container for several days. Models made with it can be painted and varnished. Air dry, or bake in a slow oven (this can take several hours). The dough can be coloured with powder paint or food colouring and glitter can even be added.

- Give each learner a piece of clay or dough. Allow them time to play with it – squashing, bending, twisting and stretching it into different shapes.
- Ask them to make a model – either of their own choice, or as directed by you – for example it could be an animal, a mask or a pot. Allow the models to dry and let the learners decorate them later.
- Give out photocopiable page 69 to the learners in pairs or small groups. Explain that they have to use the objects to try to change the shape of them by bending, squashing, stretching or twisting.
- In pairs or small groups, have the learners carry out the activities on photocopiable page 69.

Plenary

- Make a display of the finished models.
- Go over the results on photocopiable page 69.

Success criteria

Ask the learners:

- Tell me something that you can squash.
- Which can you stretch the most – a rubber band or a woolly scarf?
- Name three things that you have used today that can be twisted.
- Which items could you change the shape of in every way you tried?

Ideas for differentiation

Support: Work in a small group with these learners as they work through photocopiable page 69.

Extension: Challenge these learners to find another example of (an) object(s) that can bend, squash, stretch and / or twist.

Name: _____

Changing shape

You will need:

A rubber band, a plastic drinking straw, a piece of paper, a bath sponge and a woolly scarf.

What to do

- Try to bend, squash, stretch and twist each item.
- Complete the table to show how the shape of these things can be changed.
- Tick (✓) to show if they can be changed.

	Bend	Squash	Stretch	Twist
rubber band				
plastic straw				
piece of paper				
sponge				
woolly scarf				

- Which item could do all these things? _____

Investigating squashing and stretching

Learning objectives

● Predict what will happen before deciding what to do. (2Ep5)
● Review and explain what happened. (2Eo9)
● Know how the shapes of some materials can be changed by squashing, bending, twisting and / or stretching. (2Cc1)

Resources

Pictures (internet or books) or samples of materials, e.g. paper, fabric, soft clay, rock, wood, metal, plastic; rubber bands; bath sponges; soft foam or rubber balls; fabric pieces (can be different types); bananas; rocks (can be different types); photocopiable pages 71, 72 and 73.

Starter

• Ask the learners to think back to the previous lesson with talk partners and name some materials that it is easy to change the shape of. Discuss their responses (for example clay or dough, ourselves, bath sponge, paper, rubber band, plastic drinking straw, woolly scarf).

• If time allows, permit the learners to paint and / or decorate the models they made in the previous lesson.

Main activities

• Explain that although it is easy to change the shape of some materials, it is hard to change the shape of other materials. Ask the learners if they can think of anything that it would be hard to change the shape of. Discuss their responses.

• Show them some of the samples of materials or items available and discuss whether they would be easy or hard to change the shape of – for example soft things are easy to alter; hard things cannot easily change shape just by bending, squashing, stretching or twisting.

• Give out photocopiable pages 71 and 72 to the learners who need support and photocopiable page 73 to all the other learners. Explain that they are going to compare three different objects to see what happens to them when they are squashed and stretched. Tell the learners that they will not all be using the same objects.

• Ask the learners in pairs or small groups to carry out the activities and be prepared to tell the rest of the class what they have found out.

Plenary

• Ask the less-able learners for their conclusion (a rubber band, bath sponge and soft ball all spring back to their normal shape).

• Ask the other learners what they have found out.

Success criteria

Ask the learners:

● What is the same about a rubber band, a bath sponge and a soft ball? (They are all soft.)
● Which other material used in this lesson can behave in the same way? (Fabric.)
● What is different about the rock? (It can't do any of these things.)
● Why is this? (It is hard.)

Ideas for differentiation

Support: Give these learners photocopiable pages 71 and 72 and work with them in a small group.

Extension: Ask these learners: *Can you find another material that cannot be bent or squashed or stretched or twisted?*

Name: _____

Squashing and stretching

You will need:

A rubber band, a bath sponge and a soft ball.

What to do

● Predict what will happen when you **squash** these things.

● Use these words to help you:

get smaller   get flatter   change shape

Object	Prediction
rubber band	When I squash it, it will _____.
bath sponge	When I squash it, it will _____.
soft ball	When I squash it, it will _____.

● Predict what you think will happen when you **stretch** these things.

Object	Prediction
rubber band	When I stretch it, it will _____.
bath sponge	When I stretch it, it will _____.
soft ball	When I stretch it, it will _____.

Name: _____

Squashing and stretching

1. Write what happened after you squashed and stretched each object.

Rubber band

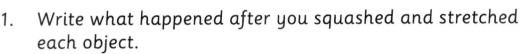

_____

_____

_____

Bath sponge

_____

_____

Soft ball

_____

_____

_____

2. Name something else that you could try to squash and stretch. Write the name of it below or draw a picture.

_____

Cambridge Primary: Ready to Go Lessons for Science Stage 2 © Hodder & Stoughton Ltd 2013

Name: _____

Squashing and stretching

You will need:

A piece of fabric, a banana or plantain, a rock.

What to do

- Predict what will happen to each of these things when you squash and stretch it.

Object	When I squash it	When I stretch it
fabric	it will _____ _____	it will _____ _____
banana	it will _____ _____	it will _____ _____
rock	it will _____ _____	it will _____ _____

- Complete the table. Tick (✓) to show what happened.

	Bend	Squash	Stretch	Twist
fabric				
banana				
rock				

Changes that last and changes that do not

Learning objectives

- Collect evidence by making observations when trying to answer a science question. (2Ep1)
- Make comparisons. (2Eo6)
- Know how the shapes of some materials can be changed by squashing, bending, twisting and / or stretching. (2Cc1)

Resources

Clay models made previously; a piece of dough or clay; balloons; springs; rubber bands; paper; coloured pencils; scissors; glue; photocopiable pages 75 and 76.

Starter

- Look at the clay models the learners made previously. Recap what the clay or dough was like before the model was left to dry. Show the learners a piece of dough or clay to remind them.
- Ask the learners to discuss with talk partners how the dough or clay has changed. Discuss their responses (it has gone hard, it is painted, it is a different shape, its shape cannot be changed, and so on).
- Recap that it is easier to change the shape of things by bending, squashing, stretching or twisting them if they are made from a soft material.

Main activities

- Blow up a balloon. Do not tie a knot in it. Ask the learners how the shape of the balloon has changed (it has stretched).
- Please be aware that some learners do not like balloons and can be frightened of them because of the squeaky noises they sometimes make. Reassure those learners and maybe sit them furthest away from you in the class while you are doing this. Ask the learners what will happen when you let the air out of the balloon (it will deflate and return to its original shape).

- With talk partners, ask the learners to think about anything else that does this (for example springs, rubber bands). Listen to their responses and demonstrate using any objects they mention if you have them available.
- Give out photocopiable page 75 to the learners who need support and photocopiable page 76 to all the other learners. Explain that they will need to look at the pictures and identify which objects will go back to their original shape after being stretched, bent, twisted or squashed. Photocopiable page 75 requires the learners to colour in, cut out and stick pictures into the correct categories on the page. The learners completing photocopiable page 76 need to draw three things that return to their original shape and three things that do not.

Plenary

- Ask the less able-learners to show and tell the rest of the class about their completed photocopiable page 75.
- Ask the other learners to share their ideas from photocopiable page 76.

Success criteria

Ask the learners:

- Tell me something that goes back to its normal shape after it has been stretched.
- Show me something that changes its shape when it is stretched.
- What happens to a balloon when it is inflated?
- What happens when the air is let out of it again?
- Why is your clay model hard now?

Ideas for differentiation

Support: Give these learners photocopiable page 75 to complete. Supervise them during the cutting and sticking activity.

Extension: Ask these learners: *Can you find or name examples of things that are made with a twist in them? (For example some types of pasta, string.)*

Name: _____

Keeping its shape

Sometimes when you change the shape of something, it stays that shape. Other things go back to their normal shape.

Colour and cut out the pictures below and stick them in the correct box.

Goes back to its normal shape	Stays as the new shape

Name: _____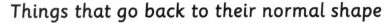

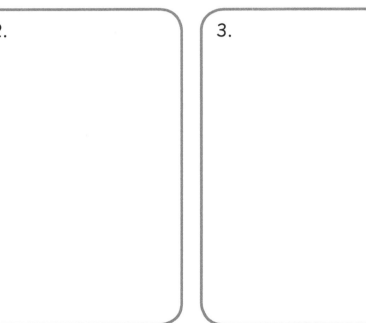

Keeping its shape

Sometimes when you change the shape of something, it stays that shape. Other things go back to their normal shape.

Draw and label three things that go back to their normal shape and three things that do not.

Things that go back to their normal shape

1.	2.	3.

Things that change shape and stay that way

1.	2.	3.

Cambridge Primary: Ready to Go Lessons for Science Stage 2 © Hodder & Stoughton Ltd 2013

Making paper aeroplanes

Learning objectives

- Use first-hand experience. (2Ep2)
- Review and explain what happened. (2Eo9)
- Know how the shapes of some materials can be changed by squashing, bending, twisting and / or stretching. (2Cc1)

Resources

Internet access; plain A4 paper; photocopiable page 78.

Starter

- Show a film clip of how to make a paper aeroplane – there are several available at www.youtube.com.
- Alternatively, if you do not have access to the internet, show the learners how to make a paper aeroplane. Follow the instructions on photocopiable page 78.

Main activities

- Give each learner a piece of paper. Display an enlarged version of photocopiable page 78 and go through the instructions step by step, clearly showing how to make the paper aeroplane.
- Move around the classroom, helping the learners to fold the paper and checking at every stage. Ensure that every learner is ready before moving on to the next step.
- Prepare some extra paper planes in case of mistakes. Some of the learners will find this tricky to do well the first time.
- It might be useful to prepare some pieces of paper with the fold lines marked on them in different colours for the learners to copy more easily. This activity has been designed so that the learners can copy you rather than having to read and follow instructions for themselves.

Plenary

- Go outside or use a large indoor space. Have an aeroplane race.
- Discuss if this is a fair test.
- Decide who the winner is and reward the champion(s).
- Explain that the paper has been bent (folded) to make a new shape. The aeroplane shape is more rigid that the original flat piece of paper. The original piece of paper can still be retrieved, but it will never return completely to its original state, due to the folding.

Success criteria

Ask the learners:

- How did we make the paper aeroplane?
- Can we get the flat piece of paper back again?
- Was this a fair test?
- Who made the best paper aeroplane?
- Why did this paper plane win the race?

Ideas for differentiation

Support: Closely assist these learners as they may find it difficult to make accurate creases and folds. Give them lots of support and encouragement at every step of the process.

Extension: Challenge these learners to make paper aeroplanes of their own designs and have a race.

How to make a paper aeroplane

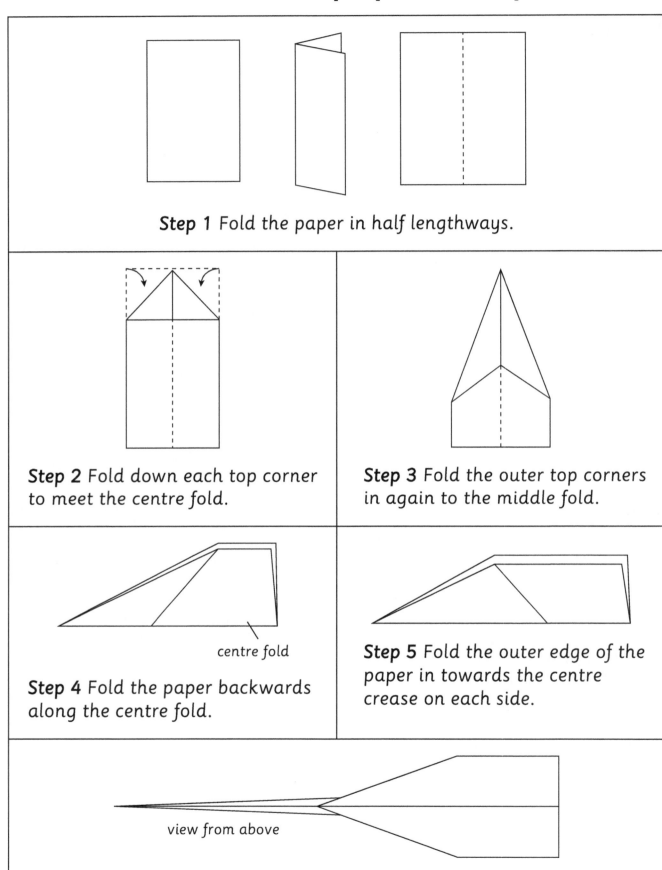

Step 1 Fold the paper in half lengthways.

Step 2 Fold down each top corner to meet the centre fold.

Step 3 Fold the outer top corners in again to the middle fold.

centre fold

Step 4 Fold the paper backwards along the centre fold.

Step 5 Fold the outer edge of the paper in towards the centre crease on each side.

view from above

Step 6 Turn it over and fold the wings out and up.

 Cambridge Primary: Ready to Go Lessons for Science Stage 2 © Hodder & Stoughton Ltd 2013

Melting

Learning objectives

- Use first-hand experience. (2Ep2)
- Talk about risks and how to avoid danger. (2Eo2)
- Explore and describe the way some everyday materials change when they are heated or cooled. (2Cc2)

Resources

Chocolate; eggs; butter; wax candles; safety matches; damp sand; a kettle full of water; ingredients for the recipe(s) on photocopiable page 80 or another recipe of your own choosing; aprons; photocopiable page 81.

Starter

- Have ready a tray containing some or all of the items listed in the resources section. Talk about each item in turn, for example:
  - Chocolate: *Do you like chocolate? When do you eat it – are there any special celebrations when we eat it?* Talk about the learners' favourite kinds of chocolate.
  - Eggs: *How do you like to eat eggs? Do you like them boiled, fried or in an omelette? Does anyone know what raw egg can be used for?* (In baking, as a glaze.)
  - Wax: Observe a burning candle. Use a tray containing damp sand to extinguish the flame and use safety matches. Ensure that the learners are a safe distance away when you light the candle. Watch it burning for a minute. Then, with talk partners, ask the learners to think of as many words as they can to describe the burning candle. Write a class list of all the words that the learners suggest.
  - Butter: *When do we eat butter? What else do we use it for?* (Cooking.)
  - Water: Boil a kettle of water, at a safe distance away from the learners. *What happens to the water when it boils?*

Main activities

- Explain that both of the things that you have demonstrated in this lesson have involved using heat – electricity from a kettle, a flame on the candle. Talk about what happens when chocolate, butter and eggs are heated (chocolate and butter melt and eggs change).
- Tell the learners that, with adult help, they are going to make something that includes melting chocolate. Share the recipe you are using with them. There are two recipes you could use on photocopiable page 80. Make sure you check for allergies. Ask the learners to wash their hands and put on aprons. Work in small groups to make the chocolate treats.
- Give out photocopiable page 81 and explain that the learners have to draw what they did in order and write a sentence for each step.

Plenary

- Let the learners eat and enjoy the food they have prepared.
- Ensure that they tidy up properly.
- Ask some of the learners to show their storyboards about how they made the recipe.

Success criteria

Ask the learners:

- How can chocolate be melted?
- Why did we break it into small pieces first?
- What happens to chocolate when it gets hot?
- What happens to it when it is put in the refrigerator?

Ideas for differentiation

Support: Supervise these learners during the cooking activity and in ordering the events for their storyboard write-ups. Alternatively, write a group storyboard with adult help.

Extension: Ask these learners to write out and decorate the recipe.

Recipes

Chocolate truffles

Ingredients

a 250g bar of chocolate

seven 15ml spoons of double cream (thick)

grated chocolate or cocoa powder

Equipment

a microwave oven or a bowl of hot water

a bowl

a grater

Method

1. Break the chocolate into small pieces.

2. An adult should melt the chocolate in the microwave or over a bowl of hot water.

3. Mix in the double cream and refrigerate for 30 minutes.

4. Roll into small balls and then roll in grated chocolate or cocoa powder.

5. Enjoy!

Chocolate skewers

Ingredients

a 250g bar of chocolate

a selection of fruits, such as banana, grapes, pineapple

Equipment

a plastic jug

a microwave oven or a bowl of hot water

knives and chopping boards

small wooden sticks

Method

1. Cut the fruit into bite-sized pieces.

2. Break the chocolate into small pieces and place in the plastic jug.

3. An adult should melt the chocolate in the microwave or over a bowl of hot water.

4. Push a small wooden stick into each piece of fruit.

5. Dip the fruit pieces in the melted chocolate and put them on a plate.

6. Let them cool.

7. Enjoy!

Cambridge Primary: Ready to Go Lessons for Science Stage 2 © Hodder & Stoughton Ltd 2013

Name: _____

Chocolate recipe

1. We made _____.

2. Use this storyboard to show how you made your chocolate treats.

   Here are some words to help you. You may not need to use them all.

   bowl   chocolate   cocoa powder   cream   fruit pieces   hot water

   jug   melt   microwave   mix   wooden sticks

1.	2.	3.

Melting materials

Learning objectives

- Predict what will happen before deciding what to do. (2Ep5)
- Make and record observations. (2Eo3)
- Explore and describe the way some everyday materials change when they are heated or cooled. (2Cc2)

Resources

Metal trays; metal spoons; plastic cups; candles or wax crayons; chocolate; ice cubes; containers to put ice cubes in; photocopiable pages 83 and 84.

Starter

- Ask the learners to remind you what happened to the chocolate in the previous lesson. Reiterate that when chocolate gets hot, it melts.
- Ask the learners to think with talk partners about how the chocolate changes. Discuss the fact that it was a solid to begin with, then it changed into a liquid when it melted, and when it cooled down it turned back into a solid again.
- Think about some other things that might melt easily, for example butter, wax, ice.
- Talk about examples of these things happening that they have seen, for example melting butter on toasted bread or on hot potatoes, wax from a candle burning, ice cubes melting in a drink.

Main activities

- Explain that in this lesson the learners are going to have the opportunity to see what happens to some things when they are left in a warm place.
- Either (according to the availability of materials) select the materials for the learners, or, if there is sufficient for them to choose, let them make their own selection. They need to choose something made of metal, something made from plastic and three things that will melt. It might be helpful to prepare trays of items labelled metal, plastic and others, that is, things that will melt easily (do not tell the learners this). Ice cubes can be kept in a vacuum flask until needed.

- Give out photocopiable pages 83 and 84 and explain that the learners need to read the instructions, set up the experiment and then look at it again later to find out what happens. Allow them to work together in pairs or small groups. They need to decide together the best place to leave their tray when they have set it up. Explain that they need to make predictions and draw before and after pictures on photocopiable page 84 to show what happens.

Plenary

- Discuss what happened to each of the items left on the tray.

Success criteria

Ask the learners:

- Did you make good predictions?
- What melted?
- Did any other items start to melt?
- Which of them have melted completely?
- Why did the metal and plastic items get warm or hot?

Ideas for differentiation

Support: Work with these learners in a small group. Help them to decide how best to arrange the objects on the tray and the best place to leave the tray during the course of the experiment.

Extension: Ask these learners: *Can we get the ice cubes back again?*

Name: _____

Melting materials 1

Try this experiment to help you find out more about melting.

You will need:

A metal tray, something made of plastic, something made of metal, three other objects – your teacher will give you some other things to choose from.

What to do

- Put your objects on the tray.
- Draw a picture of how you have set them out on the 'Melting materials 2' page.
- Predict what you think will happen. Write your predictions below.
- Leave the tray in a warm or sunny place for as long as you can.
- Look at the tray again later.
- Draw what you see now on the 'Melting materials 2' page.

My predictions

1. A plastic _____ will _____.

2. A metal _____ will _____.

3. A _____ will _____.

4. A _____ will _____.

5. A _____ will _____.

Name: _____

Melting materials 2

We left our objects on a metal tray in a warm place.

1. We put the tray _____.

2. Draw what your objects looked like at the start:

3. Draw what your objects looked like at the end:

Cambridge Primary: Ready to Go Lessons for Science Stage 2 © Hodder & Stoughton Ltd 2013

Ice-cube fishing

Learning objectives

- Use first-hand experience. (2Ep2)
- Review and explain what happened. (2Eo9)
- Explore and describe the way some everyday materials change when they are heated or cooled. (2Cc2)

Resources

Ice-cube trays, or containers for making ice in; water or liquids, e.g. fruit juice, milk; access to a freezer; a supply of ice cubes; pieces of string; aprons; towels; flat trays or containers, e.g. paint palettes, for the ice-cube challenge activity; salt; plastic cups; cold drink; the book *Matsumura's Ice Sculpture* by Anna Prokos (Longman); photocopiable page 86.

Starter

- Ask the learners to discuss with talk partners how things can be kept cool. Discuss the learners' responses (for example keep in a refrigerator, in a freezer, in ice-cold water).
- In pairs or small groups, make some ice cubes.

Main activities

- Ask the learners to think about other liquids that we can eat or drink, for example fruit juice, fizzy drinks, milk, ketchup. During this discussion, if possible, use some of these suggested liquids and foodstuffs to make ice cubes as well. Invite the learners to help you. Freeze and save these cubes for comparison later with the ice cubes made from just water.
- Go ice-cube fishing. Give out photocopiable page 86 and demonstrate to the learners how to 'fish' for ice cubes. In small groups, let them hold a competition to see who can pick up the most ice cubes.
- Tell the learners in pairs or small groups that you will give each group the same number of ice cubes, and the challenge is to see who can make the tallest tower – but they will have to be quick before the ice melts.

- Talk about how they can make the ice cubes 'stick' together (use salt).
- Hold an ice-cube tower challenge.
- Read and share the story *Matsumura's Ice Sculpture* by Anna Prokos (Longman). This is the story of a chef racing against time to make an ice sculpture.

Plenary

- Give the learners cold drinks with ice cubes made from water.
- Look at the ice cubes made from different liquids. Explain that freezing has made them solid.
- Celebrate the winning ice-cube fishers and the tower-building team.
- Explain that the string sticks to the ice because salt melts it a little bit. A pool of water forms on top of the ice and the string goes into it. The ice freezes again and the string gets trapped, so you can 'fish' the ice cube out.

Success criteria

Ask the learners:

- Have all the liquids turned into ice cubes?
- What does the salt do?
- How many ice cubes did you catch?
- How tall was the winning tower?

Ideas for differentiation

Support: Allow these learners to 'catch' an ice cube from a plate before playing the game.

Extension: Ask these learners to choose three liquids from around the room to make ice cubes with, for example paint, glue. Ask them to predict what will happen when the liquids are frozen.

Name: _____

Ice-cube fishing

You will need:

A bowl of water with ice cubes in it, salt and a piece of string (dry, hairy, thin string works best).

What to do

- Take it in turns.

- Hold the string across the top of an ice cube.

- Sprinkle some salt on it.

- The string has to be touching the ice cube.

- Count 1 elephant ... 2 elephants ... up to 10 elephants.

- Gently try to lift the string.

- Once you have each had a turn, put the ice cubes back in the bowl.

1. Now have a fishing competition! Take it in turns to 'fish' for 20 seconds. The winner is the person who 'catches' the most ice cubes.

2. I caught _____ ice cubes.

3. The winner caught _____ ice cubes.

4. What made the string stick?

_____

_____

Cambridge Primary: Ready to Go Lessons for Science Stage 2 © Hodder & Stoughton Ltd 2013

Keeping ice cubes

- Collect evidence by making observations when trying to answer a science question. (2Ep1)
- Take simple measurements. (2Eo4)
- Explore and describe the way some everyday materials change when they are heated or cooled. (2Cc2)

Resources

Internet access; ice cubes; bottles of mineral water; a metal bucket; a thermometer; salt; *Matsumara's Ice Sculpture* by Anna Prokos (Longman); plastic trays, e.g. paint palettes; fabric; bubble-wrap; paper; aluminium foil; cotton wool; a large leaf; clock or timers; photocopiable page 88; flipchart and markers or whiteboard.

Starter

- If internet access is available, go to www.bbc.co.uk and search 'bang goes the theory super cool'. This demonstrates how to make water turn into ice immediately with a simple splash or a bang. It is fun to watch.
- Alternatively, watch the clip yourself and then demonstrate it to the class.
- Talk about the previous lesson where salt made the ice melt.

Main activities

- If you did not have time to read the story *Matsumara's Ice Sculpture* by Anna Prokos in the last lesson, use it now – or re-read it. Discuss how important it is for the chef to keep the ice and stop it from melting.
- Explain to the learners that that is what they are going to find out in this lesson. Tell them that they will work in pairs or small groups. They will be given an ice cube and they have to keep it for the longest amount of time. They will be given a different material to keep their ice cube in. At the end of the lesson, they will report back on how long it took before their ice cube melted.

- Introduce the idea of fair testing, in that each group will have one ice cube and the same size piece of material. They will also use the same timers.
- Give each group a different material to work with – see the resources list. For the group working with the leaf, choose a large leaf and prepare a piece the same size as all the other materials.
- Agree how exactly the learners will set up the test. Once all the groups are ready, give out the ice cubes. Give out photocopiable page 88 for the learners to record their time on.

Plenary

- Collect each group's results and draw up a results table for the learners to copy into their own tables on photocopiable page 88.
- Talk about which was the best material for keeping an ice cube and why.

Success criteria

Ask the learners:

- Which was the best material for keeping an ice cube?
- Which ice cube melted the quickest?
- How did we make the test fair?
- Why did the ice cube last the longest in the ...?

Ideas for differentiation

Support: Assist these learners with completing the table on photocopiable page 88. Work with them in a small group, or in mixed-ability groups.

Extension: Ask these learners: *Can you find out why a vacuum flask keeps ice cubes solid?*

Name: _____

Keeping ice cubes

We tried to keep an ice cube for as long as possible.

1. Our group used _____

2. It melted after _____ minutes _____ seconds.

3. Complete the table to show each group's result.

Material	Time
fabric	
bubble wrap	
paper	
aluminium foil	
cotton wool	
leaf	

4. Which ice cube lasted the longest? _____

5. Which ice cube melted the quickest? _____

 Cambridge Primary: Ready to Go Lessons for Science Stage 2 © Hodder & Stoughton Ltd 2013

Making bread

- Use first-hand experience. (2Ep2)
- Use simple information sources. (2Ep3)
- Explore and describe the way some everyday materials change when they are heated or cooled. (2Cc2)

A raw egg; a hard-boiled egg; ingredients for baking bread – see photocopiable page 90; photocopiable page 91; aprons; cooking equipment – bowls, baking trays; jugs; kitchen scales; salt dough (recipe on page 68) or modelling clay; paints and paintbrushes; oven.

Starter

- Show the learners a raw egg. Also show them a hard-boiled egg.
- Ask the learners to compare with talk partners the difference between the two eggs.
- Ask them to name and explain some other foods that change when they are cooked, for example pasta, dough.
- Refer back to when they made the clay models / pots earlier in this unit. The dough was soft, it baked in the oven and became hard.
- Ask the learners if they know of anything we can eat that is made from dough (bread, bagels, cookies). Ask: *Do you know any of the ingredients used to make dough?* Show them the ingredients.

Main activities

- Explain that in this lesson the learners are going to make dough to bake bread. Check for any food allergies.
- They will also be able to use salt dough or clay to make a model of a plate of their favourite food and paint it.

- Give out photocopiable page 90 with the recipe on it. Go through the recipe together – display an enlarged version for reference. Demonstrate and ask the learners to help you at each stage so that they all understand how to make the dough. Show them each piece of equipment needed. Help them to measure out the ingredients carefully. (If time is at a premium, you could pre-weigh the ingredients for each group.)
- Talk about good hygiene, washing hands and equipment as part of the baking process. In small groups, with an adult helper, make bread. (Use your own recipe if you prefer.)
- Meanwhile, the other learners can make a model of a plate with their favourite meal on it, ready to paint afterwards.

Plenary

- Look at some of the model meals made by the learners. Ask them to show and tell what is on their plate and why they like to eat these things.
- Talk about making dough and why the dough was left to rise (to make the yeast work).
- Later in the day, taste the bread they have made, or allow them to take some home.

Ask the learners:

- What is dough made from?
- What is kneading?
- What does the dough look like before it goes in the oven?
- How is the bread different when it is baked?

Support: Supervise these learners closely with measuring activities during baking. Help them to write their sentences on photocopiable page 91.

Extension: Ask these learners to find out about bread that is eaten in other countries – for example baguettes in France, rye bread in Germany, chapatis in India.

Bread recipe

Ingredients

500 g strong white bread flour

$1\frac{1}{4}$ teaspoons salt

$1\frac{1}{2}$ teaspoons bread yeast

2 teaspoons sugar

300 ml water

Equipment

a large bowl

a jug

baking trays

Method

1. Ask an adult to pre-heat the oven to 230°C / 450°F.

2. Wash your hands.

3. Put the flour, yeast, sugar and salt in a large bowl.

4. Mix all the ingredients together well with your hands.

5. Slowly add the water.

6. Mix with your hands for 5 minutes.

7. Put the dough on a floured surface.

8. Take it in turns to knead the dough for 10 more minutes.

9. Shape the dough into rolls, plaits or loaves.

10. Leave the dough until it doubles in size.

11. Put the dough in the middle of the oven for 35 minutes.

12. Remove from the oven when the top of the bread is a light, golden brown.

Cambridge Primary: Ready to Go Lessons for Science Stage 2 © Hodder & Stoughton Ltd 2013

Name: _____

Baking bread

1. Draw a picture to show how you shaped the dough before baking.

2. Write a few sentences to describe what the bread looked like after baking.

    You can use some of these words to help you.

    bread   brown   crusty   golden   hard   hot   soft   solid   warm

3. When the bread was baked it was _____

    _____

    _____

Dissolving

Learning objectives

● Make and record observations. (2Eo3)
● Use first-hand experience. (2Ep2)
● Recognise that some materials can dissolve in water. (2Cc3)

Resources

Sugar; salt; sand; instant coffee; flour; oil; plastic or paper cups; water; jugs; spoons or stirrers; photocopiable pages 93 and 94.

Starter

- Ask the learners to discuss with talk partners what happens when sugar is added to a hot drink. Listen to their responses – they will probably say that the sugar disappears. At this stage it is sufficient to explain it as such. Introduce the word 'dissolve' and explain that when a solid mixes with a hot liquid, we say that the solid, in this instance sugar, has dissolved.
- Ask for suggestions of other things that they know or think might dissolve.

Main activities

- Explain that in this lesson the learners are going to test some different things to see whether they dissolve or not.
- Discuss how to make the test fair. You need to agree on the amount of water, the amount of substance and perhaps the number of stirs.
- Ask the learners to predict before they test and to tell each other what they expect will happen each time.
- Show the learners the substances available to test. You might decide to give all the groups the same substances, or ask some groups to test more substances than others in the time allowed. Differentiate this task according to the time available, the abilities of the learners and the substances to be tested.

- Organise the learners into small groups to carry out the activity. These groups can be mixed-ability or similar-ability level, depending on how you choose to structure the lesson.
- Give out photocopiable page 93 to the learners who need support and photocopiable page 94 to all the other learners. Explain that they have to complete the photocopiable page to show what they tested and what they found out.

Plenary

- Go through the results from the experiment on photocopiable pages 93 and 94.
- Explain that if a substance can still be seen in the water, we say that this substance does not dissolve.

Success criteria

Ask the learners:

● Which substances seem to have disappeared completely?
● Tell me a substance that does not dissolve.
● How can you tell if a substance has not dissolved?
● What is different about how oil behaved?
● Did you predict correctly?

Ideas for differentiation

Support: Give these learners photocopiable page 93 to work from.

Extension: Ask these learners: *How can you tell if there is sugar or salt in the water?*

Name: _____

Dissolving

When something seems to disappear in water, we call
this **dissolving**.

1. Complete the table to show what you tested and if it dissolved.

Substance	Dissolve? Tick (✓) or (✗)

2. Which substances dissolved?

    _____   _____   _____

3. Which substances did not dissolve?

    _____   _____   _____

Name: _____

Dissolving

When something seems to disappear in water, we call this **dissolving**.

1. Label the diagrams to show which things you tested.

_____   _____   _____

_____   _____   _____

2. We used _____ ml of water each time.

3. We used _____ of substance each time.

4. Complete the results table below.

Substance	Did it dissolve? (✓) or (✗)	Other changes

5. Which substances seemed to have disappeared completely?

_____

Cambridge Primary: Ready to Go Lessons for Science Stage 2 © Hodder & Stoughton Ltd 2013

Natural and man-made materials

Starter

• Give the learners in pairs or small groups a selection of materials. Ask them to sort these into natural and man-made materials.

• Discuss their groupings. (If the resources suggested are available, the groupings are as listed in the resources section.)

• Ask for a volunteer learner who does not mind wearing a blindfold. Blindfold them, then give them some objects to identify simply by touch – for example a wooden bead, a metal key. Discuss how they knew what the objects were without seeing them.

Main activities

• Show and discuss the natural materials first. Talk about where each natural material is found:
  • wax – beeswax from honeycomb, or jojoba oil
  • wool – sheep, goat, alpaca, rabbit or llama hair
  • sponges – grow under the sea
  • metal – from rocks.

These are the facts that the learners are least likely to know about some natural materials.

• Now go on to explain that man-made materials are made from natural materials. Ask the learners if any of them know what oil is used to make (plastics), or sand (glass), or iron metal (nails), or wood (paper, furniture).

• Give out magnifying glasses and allow the learners to view some hand-made paper. Ask: *Can you see the wood pulp in the paper?*

• Give out photocopiable page 96 to the learners who need support. Give photocopiable page 97 to all the other learners. Explain that they need to classify materials as natural or man-made and identify some of their uses.

Plenary

• Go over the answers to photocopiable page 96 with the learners who need support.

• Award (a) prize(s) for the materials quiz winners and good ideas from photocopiable page 97.

• Explain that natural materials can be found on or in the Earth.

• Remind the learners that man-made materials are made using natural materials to make useful products.

Name: _____

Natural or man-made?

1. Colour in the pictures.

2. Draw a red (circle) around all the things that are natural.

3. Draw a blue (circle) around all the things that are man-made.

4. Name something that is made from each of the materials below.

Material	Object
wood	
wool	
clay	
metal	

 Cambridge Primary: Ready to Go Lessons for Science Stage 2 © Hodder & Stoughton Ltd 2013

Name: _____

Materials quiz

Collect or name these things to win a prize!

Work on your own or with a partner.

You will need:

A collecting bag and a pencil.

1. Name or find something made from wax.	
2. Find an object made from wood.	
3. Name an animal other than a sheep that has wool.	
4. Find something made from clay.	
5. Find or name three different rocks.	
6. Find two different types of paper and name them.	
7. Find something made from soft plastic.	
8. Find something made from hard plastic.	
9. Name one way in which plastic and glass are the same.	
10. Find a natural material and an object made from that material.	

Unit assessment

Questions to ask

- Name a material that occurs naturally.
- What do we call materials that do not occur naturally?
- Describe two ways in which you can change the shape of something.
- What happens when chocolate gets warm or hot?
- What happens to water when it freezes?
- How can you tell if something has dissolved in water?

Summative assessment activities

Observe the learners whilst they participate in these activities. You will quickly be able to identify those who appear to be confident and those who may need additional support.

Changing shape

This activity assesses the learners' observations of how changing shape can affect dough.

You will need:

Clay or modelling dough; tools and boards; aprons; activity cards; recording pages; pencils; rulers.

What to do

- Prepare and laminate (if possible – so that they can be re-used) some activity cards that say:
  - Make a ball of clay. Squash it! Draw round it.
  - Roll a sausage shape of clay. Bend it. Draw the shape now.
  - Roll a sausage shape of clay. Twist it. Draw it.
  - Make a ball of clay. Stretch it. Measure how long it is in cm.
- Either work with a small group and instruct the learners to do the same activity at the same time, or give them each a different activity to do at the same time. Alternatively, leave the learners who do not need support to carry out the activity independently.
- Give each learner a set of labels that say 'squash', 'bend', 'twist' and 'stretch'. Ask them to label their shape and measure how long it is in centimetres.

Natural or man-made?

This activity allows the learners to classify materials as natural or man-made.

You will need:

An assorted selection of natural and man-made materials (suitable for collage making); scissors; glue; paper.

What to do

- Ask the learners to make a collage using either the natural or man-made materials available. Allow them to select what they need accordingly from all the materials available.
- Discuss their finished collage. Ask them to name some of the materials used, where they come from, some of their main uses, and so on.
- Display the finished collages around the room. Take photographs, as evidence, if required.

Written assessment

Distribute photocopiable page 99. The learners should work independently, or with the usual adult support they receive in class.

Name: _____

Heating and cooling

Use these words to help you describe what happens to these things when they are heated.

| cold | hard | hot | melt | soft |

Item	What happens when it is heated?
ice cube	_____ _____ _____
chocolate	_____ _____ _____
candle	_____ _____ _____
unbaked bread	_____ _____ _____
clay pot	_____ _____ _____

Identifying rocks

Learning objectives

- Use first-hand experience. (2Ep2)
- Make comparisons. (2Eo6)
- Recognise some types of rocks and the different uses of rocks. (2Cp1)

Resources

Flipchart and markers or interactive whiteboard; a selection of rock samples, e.g. chalk, granite, limestone, marble, sandstone, slate; labels with the name of each type of rock written on them; hand lenses or magnifying glasses; reference books or internet access; a sample of concrete or an aggregate; photocopiable pages 101 and 102.

Starter

- Ask the learners if they can remember from the previous lesson two main groups we can easily classify materials into (natural and man-made).
- Ask the learners to discuss with talk partners which natural materials come from underground. Discuss their responses and ask them to tell you where rocks come from.
- List some ways in which different rocks are used. Invite the learners to write this class list and display it prominently during the lesson.

Main activities

- Make the rock samples available to the learners. Pass them around for the learners to look at and handle. Give out hand lenses or magnifying glasses and allow them to examine the samples. Talk about different characteristics of the rocks, for example colour, particle size. Ask: *Can you see any different pieces making up the rock?* (None / large / small / tiny.) Discuss the shape of particles (round, smooth or angled) and texture.
- Write the learners' observations into a table as they make them. Use the prepared labels to label each rock sample. Discuss each sample in turn, using words as used by the learners when making their observations.

- Give the learners in pairs or groups some smaller rock samples to identify. Give the learners who need support photocopiable page 101. Tell these learners that they need to draw and name three different rocks. Give all the other learners photocopiable page 102. Explain that they have to complete the table for the rock samples you give them. (If there are plenty of rock samples, allow the learners to choose; if so, limit / vary choices by pre-grouping the rocks into sedimentary, igneous and metamorphic trays, and asking the learners to choose at least one from each tray.)

Plenary

- Look at the drawings of rocks made by the learners who have completed photocopiable page 101.
- Ask some of the learners who completed photocopiable page 102 to share some of their findings.
- Show the learners a piece of concrete or an aggregate. Explain that this is an example of a man-made rock. Discuss how it is similar to natural rock.

Success criteria

Ask the learners:

- Name a type of rock that you have looked at in this lesson.
- Show or tell me the name of a rock that is smooth.
- Find me a rock sample with smooth particles in it.
- Can you give me an example of an igneous / metamorphic / sedimentary rock?

Ideas for differentiation

Support: Give these learners photocopiable page 101 to work from.

Extension: Using reference books or the internet, if available, ask these learners: *Can you name an example of an igneous, a metamorphic and a sedimentary rock?* (For example pumice, marble, sandstone.)

Name: _____

Looking at rocks

1. Use a magnifying glass to look at three different rocks.

2. Name them and draw them below.

1. _____

2. _____

3. _____

Name: _____

Looking at rocks

1. Use a hand lens or magnifying glass to look at your rock samples.

2. Complete the table below. You can use the class table to help you with this.

Rock	Colour	Shape of particles	Size of particles	Texture
1.				
2.				
3.				
4.				
5.				
6.				

3. Write one use for any of the rocks you have looked at.

_____ is used for _____

_____

Cambridge Primary: Ready to Go Lessons for Science Stage 2 © Hodder & Stoughton Ltd 2013

What are rocks used for?

- Use first-hand experience. (2Ep2)
- Make and record observations. (2Eo3)
- Recognise some types of rocks and the uses of different rocks. (2Cp1)

Pictures (internet or books) of different types of rock or rock samples as before; photocopiable page 104; digital camera; clipboards; pencils.

Starter

- Play a game to identify some rock samples. Describe a rock and ask a learner to find it from the pictures or samples, for example rough, yellow, small grains that can easily be rubbed off – sandstone. Repeat this several times for different rocks.
- Ask one learner to find the sample, and the rest of the class to help them name it.

Main activities

- Explain that you are going to take the class for a walk around school. This will be an opportunity to look for different types of rocks and how and where they are used. Give out photocopiable page 104 for the learners to complete as they go around. (The photocopiable pages could be given one between two, or one learner could act as scribe for a small group.) Tell them that they might not be able to write something in every column for every rock that they see.
- Go for a walk, taking digital photographs of anything of particular interest. Allow the learners to take photographs too. Stop at particular points and guide the learners to look up high (roofs), down at the ground (road surfaces) and the buildings themselves.

- Return to the classroom and discuss the learners' findings. Look at any photographs that have been taken. Examples that might be seen include:
  - granite – hard and strong, used for road surfaces and some buildings
  - marble – usually used for decorative purposes such as statues
  - sandstone – occurs in blocks and is often used for paving
  - slate – often used for roof tiles in some parts of the world because it is thin.
- Ask the learners to bring in information and pictures of examples of different rocks being used around them. Over time, build up a classroom display of these. Supplement it with the learners' work from this chapter.

Plenary

- Play the Starter game again, identifying rocks – but allow the learners to describe the rocks this time.
- Discuss why particular rocks are used for specific purposes, for example marble for statues, and so on.
- Discuss the building materials that may have been used in the construction of the school buildings and / or the learners' homes.

Ask the learners:

- What rocks are most of the buildings around school made from?
- What is the roof covering made of?
- How are the roads and paths near school constructed?
- Are there any special places that use different rocks?

Support: Work with these learners as a small group and act as scribe for the group on the walk.

Extension: Ask these learners to find out about any statues nearby or famous in history.

Name: _____

Rock hunt

Complete this table with a friend or for your group as you walk around school.

(You might not be able to write something in every column.)

Rock	Appearance	Place

 Cambridge Primary: Ready to Go Lessons for Science Stage 2 © Hodder & Stoughton Ltd 2013

Testing for wear

- Recognise that a test or comparison may be unfair. (2Ep6)
- Review and explain what happened. (2Eo9)
- Recognise some types of rocks and the uses of different rocks. (2Cp1)

Pictures (books or internet) of weathered statues, e.g. the Sphinx; rock samples; magnifying glasses or hand lenses; sandpaper; safety goggles; photocopiable page 106; coloured pencils.

Starter

- Look at or show some pictures of weathered statues or buildings. Have the learners discuss with talk partners what has happened over time. Think about any examples that might have been seen on the walk around school in the previous lesson.
- Listen to the learners' responses and ascertain from them why these changes might have happened (effects of weathering, rain and wind).
- Talk about the rocks that these buildings and statues have or might have been made from.

Main activities

- Explain to the learners that in this lesson they are going to test rock samples to see how easily they wear away, that is, get worn down. Tell them that this will involve rubbing pieces of rock to see what happens to them. They will need to look at the rocks before and after and compare them to see if there is any change.
- Work together as a class to agree how to make the test fair.

- Tell the learners that they will use a piece of sandpaper each time. Ensure that the pieces of sandpaper are the same size. Show them a piece of sandpaper and ask them if it is fair to use the same size pieces. Also try to ensure that the rock samples are of a similar size. Agree whether a new piece of sandpaper should be used each time, the number of times each sample will be rubbed, the place on the rock to be rubbed and how to make sure that the same pressure is applied each time (let the same learner do the rubbing in each group).
- It is important for the learners to begin to understand that only one thing should change in a fair test. In this test it should just be the rock sample that is different each time.
- Give out the rock samples and photocopiable page 106 for recording purposes. Explain to the learners that they will need to draw the rocks before and after the rubbing test.

Plenary

- Go over the learners' responses to the test.
- Discuss everyday uses for each of the rocks tested.
- Verify that the learners are able to compare rocks by how they wear.

Ask the learners:

- Which rock wore away easily?
- Which rock did not seem to change?
- What is sandstone usually used for?
- Why is marble a good material to make statues from?

Support: Work in a small group with these learners, or organise them to work in mixed-ability groups in the class.

Extension: Ask these learners to rank all the materials tested from best to worst wear and to tell the rest of the class why.

Name: _____

Testing rocks for wear

1. Record what each rock looks like before and after rubbing it with sandpaper.

Rock	Before	After

2. Which rock was the easiest to wear away? _____

3. Which rock did not wear away? _____

Cambridge Primary: Ready to Go Lessons for Science Stage 2 © Hodder & Stoughton Ltd 2013

Are rocks waterproof?

- Make comparisons. (2Eo6)
- Review and explain what happened. (2Eo9)
- Recognise some types of rocks and the uses of different rocks. (2Cp1)

Rock samples; water; pipettes or droppers; containers to put rocks in; information sources (books or internet); photocopiable pages 108 and 109.

Starter

- Refer back to the previous lesson where you talked about weathering. Weathering is usually a result of rain or wind damage. Tell the learners that because rocks are made from different sizes and shapes of particles, the way they behave when water touches them is also different.
- Look again at pictures of weathered buildings or statues. Remind the learners of the damage that moisture or wet can cause.

Main activities

- Explain that in this lesson the learners are going to test how well or badly water can get into rocks. Give out photocopiable page 108 to the learners who need support. Tell them that they need to follow the instructions carefully in order to carry out the test.
- Give photocopiable page 109 to all the other learners. Have them work in pairs or small groups. Tell these learners before they do the test and during the test that you will check to see if they have made it a fair test. Remind them that they need to think about how much water to add each time. They will also need to agree where to put the water drops each time on the rock sample.

- The challenge for them is to put the rocks that they test in order from most to least waterproof. (Remind them that waterproof means that it does not let water in.) Move around the class while they carry out the test, offering support, advice and encouragement.

Plenary

- Ask the learners who needed support what they found out about which rocks absorb water and which do not.
- Ask some of the other learners to share their results with the rest of the class.
- Do they agree about which rocks are the most and least waterproof from those tested?
- Introduce the particular scientific vocabulary of 'permeable' (lets water through) and 'impermeable' (doesn't let water through). Learners like to hear and repeat long scientific words. They should not be expected to remember these terms at this stage.

Ask the learners:

- Tell me a rock that soaks up a lot of water.
- Which rocks do not soak up much or any water?
- Which rock soaked up water quickest?
- Which rocks are waterproof?

Support: Give these learners photocopiable page 108 to complete.

Extension: Ask these learners if they can find out a single scientific word that means 'lets water through' (permeable).

Name: _____

Are rocks waterproof?

You will need:

- rock samples

- water

- a pipette or dropper

What to do

- Drop five drops of water on to the rock sample.
- Wait to see if the water soaks into the rock.
- Record what happens in the table below.

Rock	Water soaks in?	Water stays on top?

1. Which rock is best for not soaking water up? _____

2. Which rock soaked up a lot of water? _____

 Cambridge Primary: Ready to Go Lessons for Science Stage 2 © Hodder & Stoughton Ltd 2013

Name: _____

Are rocks waterproof?

You will need:

- rock samples

- water

- a pipette or dropper

What we did

1. _____

2. _____

3. _____

What happened?

Put the rocks in order below from 1 to 6, starting with the most waterproof.

	Rock	How could you tell?
most waterproof	1.	
	2.	
	3.	
	4.	
	5.	
least waterproof	6.	

Making edible rocks

Learning objectives

- Use a variety of ways to tell others what happened. (2Eo5)
- Make comparisons. (2Eo6)
- Recognise some types of rocks and the uses of different rocks. (2Cp1)

Resources

Pictures (internet or books) of different types of rock and of their formation – sedimentary, igneous and metamorphic; photocopiable page 111; milk, white and dark chocolate; cling film; plastic or paper cups; teaspoons; plastic knives.

Starter

- Show or look at examples of each type of rock – igneous, metamorphic and sedimentary. Refer to rocks that the learners have used in recent lessons. If possible, show film clips of how these rocks are formed.

Main activities

- Explain that in this lesson the learners will have the opportunity to make some rock from chocolate to see how the different types of rock are formed. Give out photocopiable page 111, which gives instructions for how to make sedimentary and metamorphic rocks from chocolate.

- Demonstrate and allow the learners to follow the instructions and to copy you making each type of rock. Explain that sedimentary rock is made from layers of rock being built up under the sea. Examples of sedimentary rocks include sandstone, limestone, chalk and shale.

- Explain that metamorphic rock is rock that has been changed by heat and pressure underground inside the Earth, for example marble or slate.

- Explain that igneous rocks such as granite or basalt are forced up out of the ground by volcanoes.

- Demonstrate making igneous rock from chocolate: Have a ready-prepared sample of sedimentary rock. Put it into a square of cling film, then roll it into a ball. Dip it (wrapped in clingfilm) into a cup of hot water for 30 seconds. Lift it out and leave it to cool. Unwrap it and then cut through the ball.

Plenary

- Discuss what you see each time; the sedimentary rock is made up of layers, the metamorphic rock has the layers mixed up and the igneous rock has melted together.

- Eat the rocks! (Check for any food intolerances or allergies amongst the learners.)

- Look at samples of each type of rock – igneous, metamorphic and sedimentary. Use the relevant terms, but the learners should not be expected to remember these terms at this stage.

Success criteria

Ask the learners:

- How is sedimentary rock made?
- What changes turn rock into metamorphic rock? (Heat and pressure.)
- Where does igneous rock come from? (Volcanoes.)
- Name a sedimentary / igneous / metamorphic rock.

Ideas for differentiation

Support: Supervise these learners while making the edible rocks. Ensure that they can clearly see the adult helper who is demonstrating what to do, or invite these learners to make it with you step by step.

Extension: Ask these learners: *What is the liquid rock that comes out of a volcano called?*

Name: _____

Rocks you can eat!

Sedimentary rock

You will need:

Grated milk or dark chocolate, grated white chocolate, a plastic or paper cup, a square of cling film.

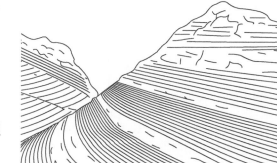

What to do

- Put the cling film into the cup.

- Put two to three teaspoons of grated milk or dark chocolate into the cup.

- Place two to three teaspoons of grated white chocolate into the cup.

- Add another layer of milk or dark chocolate.

- Fold the cling film over the top.

- Push down hard with your fingers until the chocolate feels like it sticks together.

- Pull it out gently and unwrap it.

- Break the rock in half.

What can you see? _____

Metamorphic rock

- Wrap one half of your sedimentary rock in cling film.

- Squeeze it into a ball and try to keep it in a ball shape.

- Roll it around and keep squeezing it for a few minutes.

- Let it cool for two minutes.

- Unwrap it and cut the ball in half.

What can you see? _____

Different kinds of soil

Learning objectives

- Use first-hand experience. (2Ep2)
- Make and record observations. (2Eo3)
- Recognise some types of rocks and the uses of different rocks. (2Cp1)

Resources

Flipchart and markers or whiteboard; a sample of rock; some crushed rock; a plastic cup full of soil; a plastic cup full of water; a dead plant; photocopiable pages 113, 114 and 115; samples of different soils, e.g. sandy, clay, peat-free compost (do not use peat as it is environmentally sensitive); hand lenses or magnifying glasses; flat trays, e.g. paint palettes; water; containers to mix soil with water in; books or internet for sources of information; sieves.

Starter

- Ask the learners to discuss with talk partners what soil is. Listen to their responses and write down a list of all the things that they think soil includes.
- Explain that really, soil is a mixture of broken-up rocks, mixed with plants that have died or are decaying. This mixture is called humus and it can hold water and stick the rock particles and dead plant parts together in clumps.
- Make some soil from these things – to the crushed rock, add the water and plant material. This is your soil. Give out photocopiable page 113 and ask the learners to complete the page, showing how to make soil.

Main activities

- Explain that during this lesson the learners will look at different kinds of soil. Tell them that they will look at and compare different samples.
- Either give different groups one kind of soil each to look at and then ask each group in turn to feed back to the rest of the class about the type of soil they have looked at, or give all groups the same, different types of soil to look at several examples.

- Give out photocopiable page 114 or 115. Photocopiable page 114 is for recording observations about one type of soil. Photocopiable page 115 is for recording observations about several different types of soil, or for collating whole-class results during the Plenary.
- Demonstrate sieving to separate particles of different sizes and mixing with water to see if it sticks together in clumps. Remind the learners to use magnifying glasses or hand lenses to look at particle size and shape.

Plenary

- Discuss the characteristics of each type of soil, referring to the tables on photocopiable pages 114 and 115.
- Explain that soil is the top layer of rock on the Earth's crust.

Success criteria

Ask the learners:

- What does soil contain?
- Which types of soil can hold a lot of water?
- Which soil does water drain through easily?
- Which type of soil would be best to grow plants in?

Ideas for differentiation

Support: Give these learners photocopiable page 114 or limit the number of soils they look at on photocopiable page 115.

Extension: Challenge these learners to find out what the structure of the Earth is like from core to crust. Use reference sources.

Name: _____

How to make soil

1. Write and draw, in order, to show how you made soil.

2. Use these words to help:

> crushed   plant   rock   soil   water

1.	2.
_____ _____ _____	_____ _____ _____
3.	4.
_____ _____ _____	_____ _____ _____

Name: _____

Looking at soil

1. (Circle) the type of soil you looked at.

| chalky | clay | loam | peat | sand | silt |

2. Complete the table below to describe what this type of soil is like.

Colour	
Texture	
Particles (sieve to separate) What size are they? What shape are they?	
What happens when the soil is mixed with water?	

3. Do you think that this would be a good soil to grow plants in? yes / no

 Cambridge Primary: Ready to Go Lessons for Science Stage 2 © Hodder & Stoughton Ltd 2013

Name: _____

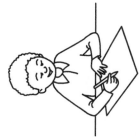

Looking at different soils

1. Complete the table below for the soils you have looked at.

Type of soil	Colour	Texture	Size of particles	Size of particles (find by sieving)	What happens when it is mixed with water?

2. Which type of soil would you choose to plant seeds in? _____

What is in soil?

Learning objectives

● Collect evidence by making observations when trying to answer a science question. (2Ep1)

● Make and record observations. (2Eo3)

● Recognise some types of rocks and the uses of different rocks. (2Cp1)

Resources

Chart showing the different layers in the Earth's crust; soil samples; plastic beakers; water; teaspoons; photocopiable page 117; poster paper; collage-making materials; scissors; glue.

Starter

• Ask the learners who did the extended activity from the previous lesson to tell the rest of the class about the structure of the Earth from core to crust.

• Prepare a chart showing the different layers within the Earth's crust, namely ground level, top soil (contains decomposed remains of living material, plants and fine rocks), subsoil (larger pieces of rock, less plant remains) and bedrock (solid rock). Alternatively, use pictures (from books or the internet). Use these terms, but do not expect the learners to use and remember them.

Main activities

• Explain that in this lesson the learners are going to look at different soils to see the particles and colours in each.

• Give out photocopiable page 117 and tell the learners to read and follow the instructions to carry out the test. Leave the samples to settle over several days in order to see better results, but the learners will be able to record their results within the usual lesson time.

• Different layers should be seen settling out as the water begins to clear. Heavier particles such as small stones will settle at the bottom of the beaker. Above this there might be a sand layer. Above the sand layer there could be a layer of silt. On top of the silt layer will be the clay layer, and on the surface of the water you should be able to see humus floating.

• Explain that darker soils tend to contain more humus. Sandy soils tend to be pale and gritty. Red-coloured soils are usually clay based and contain quite fine particles.

Plenary

• Compare the soil samples in terms of particle layers and colour.

• Identify the particular layers in each sample.

• Ensure that the learners know that soils contain different particles and can be different colours.

Success criteria

Ask the learners:

● Which type of soil did you use?

● How many layers can you see?

● Is anything floating on top of the water?

● What has sunk to the bottom of the beaker?

● What do you notice over time?

Ideas for differentiation

Support: Work with these learners in a small group and guide their observations by asking relevant questions (as used in the Plenary).

Extension: Have these learners make a poster or collage showing all the different layers. Ask them to show and display their poster or collage when it is finished. This could be a small-group activity.

Name: _____

Separating soil

You will need:

A soil sample, a plastic beaker, water and a teaspoon.

What to do

- Put three spoons of soil in the beaker.
- Fill the beaker with water.
- Stir.
- Leave to settle.

1.   What can you see?

2.   Draw and label the layers below.

Which type of soil holds most water?

● Recognise that a test or comparison may be unfair. (2Ep6)
● Make suggestions for collecting evidence. (2Eo1)
● Recognise some types of rocks and the uses of different rocks. (2Cp1)

Internet access; pictures (from books or the internet) of rainy seasons around the world and rainy weather conditions; soil samples; clay; sand; water; plastic beakers or cups; funnels; filter paper; plastic trays, e.g. paint palettes; photocopiable pages 119 and 120; a variety of timers, e.g. sand timers, stopwatches.

Starter

• Look at the range of pictures available. As a class, talk about where in the world these events might be happening.
• Ask: *Have you had any experience of flood or monsoon conditions?* Ask the learners to share their experiences with the rest of the class.
• If internet access is available, show film clips of flood or monsoon conditions (search at www.youtube.com).

Main activities

• Explain that in this lesson the learners are going to find out more about how water drains through different soils. Describe it as similar to the activity where they compared water going through rocks.
• Demonstrate how to use the filter funnel and filter paper. If laboratory equipment is not available, paper towels can be substituted for filter paper and the tops of plastic drinks bottles can be cut off and inverted to make funnels. Plastic or paper cups can also be used instead of plastic beakers. Demonstrate using the equipment that the learners will use. Use sand and clay when carrying out the demonstration and compare what happens. Try to filter sand and water and compare it with clay and water.

• Discuss how to make this a fair test – use the same amount of soil and water each time. Ask: *How will you measure the amounts of water and soil?*
• Give out photocopiable pages 119 and 120. Explain that the learners will need to carry out their test as fairly as possible and to use their photocopiable pages to write about what they did. Explain that they can complete each section and then show you before continuing. This may take one or more lessons.
• Guide discussions about whether they will time how long the water takes to drip through or measure the amount of water that drips through in a given time. Either method is acceptable, but this might depend on how capable the learners are at using any available timers.

Plenary

• Find out if timing or measuring the amount of water gives the same results.
• Agree as a class which is the best and which is the worst type of soil for draining water through.

Ask the learners:

● How much water and soil did you use each time?
● How did you measure the results?
● Which soil does water go through quickest?
● What could you do to make water drain away more easily?

Support: Complete the investigation pages step by step with these learners.

Extension: Ask these learners: *What could you do to make water drain away more easily?*

Name: _____

My investigation 1

Question: _____?

Predict: I think that _____

_____ will happen.

 What to do

We will need (write a list or draw)

Name: _____

My investigation 2

 What we did

 What happened

The answer to my question is:

_____
_____
_____
_____

Cambridge Primary: Ready to Go Lessons for Science Stage 2 © Hodder & Stoughton Ltd 2013

Which type of soil do seeds grow best in?

Learning objectives

- Predict what will happen before deciding what to do. (2Ep5)
- Talk about predictions (orally and in text), the outcome and why this happened. (2Eo8)
- Recognise some types of rocks and the uses of different rocks. (2Cp1)

Resources

Internet access; seeds; soil samples; water; containers for growing seeds in; photocopiable pages 122 and 123.

Starter

- Ask the learners to think with talk partners about what plants need to grow well.
- Discuss the learners' responses (water, air, Sun, soil).
- If internet access is available, go to www.scibermonkey.org: ages 5 to 7 → Plants and animals → Growing plants. Do the activity and the quiz, if time. This will reinforce for the learners the things that plants need to grow well.

Main activities

- Explain that in this lesson the learners will plant some seeds in two different types of soil. They have to choose the soils that they think will be best. They might decide to choose a good and a bad soil, to compare their results. It is up to them.
- Organise the learners into groups for this activity. They can work in mixed-ability or similar-ability groups.
- Show the learners the seeds that are available – just have one kind. Perhaps show them a picture from the seed packet of what the grown plant should look like. Choose quick-growing seeds, for example peas, beans or sunflowers.

- Explain that the learners have to decide how many seeds, the amount of water and the soil to use each time. Ask: *How often will you water the plants? Where will you leave the plants to grow?* Give the learners some thinking, planning and discussion time.
- Have them set up and carry out the activity.
- Give out photocopiable page 122 for the learners to complete now and photocopiable page 123 to complete over time.

Plenary

- Discuss and find out if all the groups planned a fair test.
- Reiterate that to be completely fair, the only thing that should be different between the two pots is the soil.
- Explain that the learners will need to look at how the seeds are growing over the next few days and weeks.

Success criteria

Ask the learners:

- How many seeds did you plant in each pot?
- How much soil did you use?
- How much water did you add?
- Which two different types of soil did you choose?
- Where did you put the pots?
- What do you predict will happen?

Ideas for differentiation

Support: Assist these learners with the design of their fair test. Ensure that they have a similar approach to planting seeds in two different soils.

Extension: Using different types of soil, get these learners to mix them with water and use them as natural paints, then paint a picture of a garden.

Name: _____

Which type of soil do seeds grow best in?

You will need:

Two seeds, two pots, water, two different types of soil.

What to do

- Make it a fair test.
- Plant the seeds.
- Label the pots.
- Water the seeds.

1. Draw a picture in the box to show what you did.

2. I planted (what type?) _____ seeds.

3. I used _____ soil and _____ soil.

4. I used _____ of soil each time.

5. I used _____ of water each time.

6. Which plant will grow best? I think that the plant growing

   in the _____ soil will grow best.

Cambridge Primary: Ready to Go Lessons for Science Stage 2 © Hodder & Stoughton Ltd 2013

Name: _____

Which type of soil do seeds grow best in?

1. Choose two different types of soil and plant your seeds.

2. Draw pictures to show how the plants grow over time.

Day	Soil 1 is _____	Soil 2 is _____

3. Which seed grew best and why?

_____

_____

Rocks and soils – what do you know?

- Ask questions and suggest ways to answer them. (2Ep4)
- Use first-hand experience. (2Ep2)
- Recognise some types of rocks and the uses of different rocks. (2Cp1)

Resources

Pictures (books or internet) of precious stones; pebbles; jewellery – real or imitation; photocopiable pages 125, 126, 127, 128 and 129.

Starter

- Look at the pictures of precious stones and ask the learners to name any, for example diamond, ruby, emerald, sapphire. Talk about colours of popular gemstones, for example diamonds – colourless; rubies – red; emeralds – green; and sapphires – blue.
- With talk partners, think about special occasions when jewels are given as gifts, for example weddings, Divali, from kings and queens.
- Look at pictures of famous jewels such as the Agra diamond, the Koh-I-Noor diamond – jewels of national interest. Tell the stories behind how these gifts were given or obtained and where they were discovered.
- Pass round some items of jewellery – real or imitation. Ask the learners to identify the occasions they might be worn and to identify any of the precious stones in them.

Main activities

- Explain that rocks sometimes contain special particles that turn into precious stones when the rock gets heated or pressed very hard while in the ground over hundreds of years. They are found when rock is quarried, or mined, for building. Quarries and mines are places where rocks are brought out of the ground.
- Ask the learners if they know what very large pieces of rock are called (boulders).

- Remind the learners that soil is really lots of pieces of broken-up rocks. Show them some pebbles and ask what they are. Ask: *Where are pebbles usually found? What makes them round or smooth?*
- Explain that this is the end of this unit of work about rocks and soils and so you are going to give them a quiz to find out how much they can remember.
- Split the class into at least two teams of mixed ability, or allow the learners to choose teams. Either ask the questions to each team in turn, or give each team an answer page and ask all teams all the questions. Mark the answers at the end to find the winners.
- Copy and laminate (if possible) the questions on photocopiable pages 125, 126 and 127 to re-use in future. The learners could be chosen to ask a question each time.

Plenary

- Mark the quiz together, or mark it separately and announce the winning team.
- Award (a) small prize(s) for the winning learners.
- Deal with any misconceptions that arise.

Success criteria

Ask the learners:

- How many did the winning team get right?
- Which questions did you get wrong?
- Were there any questions that were confusing for you to answer?
- Did the team disagree on any answers given?

Ideas for differentiation

Support: Organise these learners into mixed-ability teams with their classmates.

Extension: Ask these learners questions (for example 2 and 4) that have not been specifically taught during this unit.

Rocks and soils quiz 1

1. Which rock are lots of statues made from?

2. Name a rock that can float in water.

3. What do you call a place where rocks can be dug out of the ground?

4. Name a rock that can be split into very thin layers.

5. Name a rock that has grainy particles in it.

6. Name a rock that has crystals in it.

Rocks and soils quiz 2

7. Which type of soil floods after lots of rain?

8. What do we call the best type of soil for growing plants in the garden?

9. Which type of soil drains very quickly after lots of rain?

10. What is another name for giant rocks?

11. What do we call small rounded stones?

12. Name a rock that can soak up water.

Rocks and soils quiz 3

13. Which rock is used to make glass?

14. Is glass natural or man-made?

15. What colour are rubies?

16. What is another name for precious stones?

17. What colour are sapphires?

18. Name a precious stone that is colourless.

Name: _____

Quiz team page

Team name: _____

1.	2.
3.	4.
5.	6.
7.	8.
9.	10.
11.	12.
13.	14.
15.	16.
17.	18.

Cambridge Primary: Ready to Go Lessons for Science Stage 2 © Hodder & Stoughton Ltd 2013

Teacher's answer page

1. Marble	2. Pumice
3. A quarry or mine	4. Slate
5. Any of these: limestone, sandstone, chalk	6. Either of these: marble, granite
7. Clay soil	8. Loam
9. Sandy soil	10. Boulders
11. Pebbles	12. Any of these: limestone, chalk, sandstone
13. Sand or sandstone	14. Man-made
15. Red	16. Jewels or gemstones
17. Blue	18. Diamond

Unit assessment

Questions to ask

- Name some uses of rocks.
- Which rock is sometimes used to make statues?
- Name a soft rock.
- Name a rock that is hard.
- Which rocks can be used for building houses?
- What is soil?

Summative assessment activities

Observe the learners whilst they participate in these activities. You will quickly be able to identify those who appear to be confident and those who may need additional support.

Looking at rocks

This activity enables the learners to classify different rocks.

You will need:

A selection of different rock samples, previously seen and used by the learners.

What to do

- Ask the learners independently to group and re-group the different rocks.
- Ask them to justify the reasons for any groupings that they make.
- Ask pertinent questions such as: *What is the name of a particular rock? What is this rock used for?*
- Record the learners' responses on a class checklist or in photographic evidence, if this would be more useful to you.

All about rocks

This activity permits you to question the learners' knowledge and understanding about rocks.

You will need:

A selection of pictures of rocks showing fossils, lava from a volcano, rock faces in nature, a mine, and so on – that is, pictures to use as a stimulus for a discussion about rocks.

What to do

- Working with a small group or in a one-to-one situation with individual learners (time permitting), ask them to select a picture or pictures and to tell you about what the pictures tell us about rocks. Use guided questioning to draw out their responses.
- On the basis of their responses, you should be able to gauge the depth of their knowledge and understanding about rocks covered in this unit.

Written assessment

Distribute photocopiable page 131. The learners should work independently, or with the usual adult support they receive in class.

Name: _____

Looking for rocks

Circle as many uses of rocks as you can find in this picture.

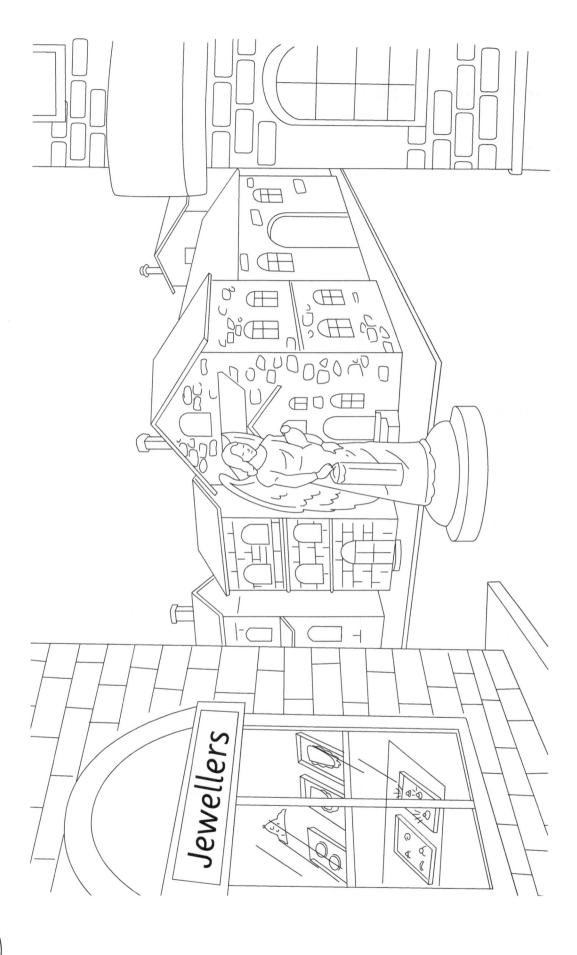

Jewellers

Light and the Sun

Learning objectives

- Use simple information sources. (2Ep3)
- Make comparisons. (2Eo6)
- Explore how the Sun *appears* to move during the day and how shadows change. (2Pb1)

Resources

Flipchart and markers or whiteboard; pictures (from books or the internet) of the Sun and various light sources; photocopiable pages 133 and 134.

Starter

- Begin by explaining to the learners that what they will learn in this chapter builds on the work they did on light and dark in Unit 1A: 2.1 Light and dark.
- Ask the learners to discuss with talk partners facts that they already know and can remember about the Sun.
- Write (or ask the learners to write) their ideas on the flipchart or interactive whiteboard as a concept map. (This is simply a chart with the word 'Sun' written in the centre of the piece of paper. Contributions are added in writing or by drawing as reminders of what 'Sun' prompts the learners to think about when they hear that word. Label lines can be drawn radiating away from the central word; these can be straight or wavy.)
- Prompt the learners' thinking to encourage them to make links between what is written on the concept map. Join these links by adding lines between words that can easily be associated together. This will give you an indication of what the learners have remembered well and what they have not and if they can link facts logically.
- Discuss their responses, for example the Sun is the Earth's light source, darkness is the absence of light, the Sun is the reason we have day and night. Encourage as many learners as possible to contribute to this discussion.

Main activities

- Explain to the learners that in this lesson they are going to think again about light sources.
- Give out photocopiable page 133 to the learners who need support and photocopiable page 134 to all the other learners. Explain that the learners completing photocopiable page 133 need to identify and name different light sources, and photocopiable page 134 requires the learners to draw and name different light sources and their fuel.

Plenary

- Go over the learners' responses to photocopiable page 133. Classify all the suggested light sources as natural or artificial.
- Go over the learners' responses to photocopiable page 134 and discuss the source of energy each time, for example electricity, wax, wood, chemicals inside the animal, hot gas.

Success criteria

Ask the learners:

- What is a light source?
- Name a natural source of light.
- What is the light source for a torch?
- What makes the Sun glow?
- Tell me something that gives a natural flame.

Ideas for differentiation

Support: Give these learners photocopiable page 133 to complete.

Extension: Ask these learners to label each energy source on photocopiable page 134 as natural or artificial.

Name: _____

Light sources

1. Use these words to name the pictures showing different sources of light.

> camp fire   Divali lamp   glow-worm   lamp   Sun   torch

_____   _____

_____   _____

_____   _____

2. Name a natural source of light. _____

3. Name a man-made source of light. _____

Name: _____

Light sources

Draw and name five different sources of light in the table below. Use your class concept map to help you.

Light source	Source of energy
candle	wax

Cambridge Primary: Ready to Go Lessons for Science Stage 2 © Hodder & Stoughton Ltd 2013

Making shadows 2

- Use first-hand experience. (2Ep2)
- Identify simple patterns and associations. (2Eo7)
- Explore how the Sun *appears* to move during the day and how shadows change. (2Pb1)

Torches; a selection of objects for making shadows; flipchart and markers or whiteboard; a screen; photocopiable pages 136 and 137.

Starter

- Play a game by using a familiar object to make a shadow on the screen and asking the learners to guess what is making the shadow, for example a key or a ball. Ask: *How could you tell what it is?* (By the shape.)
- Ask the learners to think with talk partners of as many ways as they can to describe a shadow. Listen to their responses and write a list of 'shadow facts' on the flipchart or interactive whiteboard. Suggestions from the learners might include such things as shadows can change shape, move, are black or dark coloured, follow you around, need light.

Main activities

- Ask the learners to tell you how the shadow was formed when they played the guessing game in the Starter activity. (This will tell you if they have remembered that light is blocked and cannot go through the object, so the shadow is formed.) If it becomes clear that they have not remembered from before, ensure that you remind them about how shadows are made.
- Invite some of the learners to come up and make a shadow with a different object of their own choosing. Ask questions such as: *Why is there a shadow? What is making the shadow? What shape is the shadow?* This type of questioning will focus the learners on what is making the shadow.

- Explain that in this lesson they are going to make and change shadows. Give out photocopiable pages 136 and 137 and explain that they will make shadows, try to make shadows get bigger and smaller or even change the shape of shadows.
- Organise the learners into pairs or small groups to carry out this activity. Mixed-ability groups would work well in this instance, giving an opportunity for the less-confident learners to work alongside their friends.

Plenary

- Invite each different group to demonstrate to the rest of the class how they made their shadows, and how they made them bigger or smaller.
- Ask if any of them managed to change the shape of a shadow. Discuss why this was impossible (the shadow is always the same shape as the object that makes it – this is a difficult concept for some young learners to understand).

Ask the learners:

- How did you make your shadow?
- What do you do to make a shadow smaller?
- What happens when you move the object closer to the torch?
- Why do shadows form?

Support: Assist these learners with the completion of photocopiable page 137.

Extension: Ask these learners: *Can you make the shadow look as if it is moving away into the distance, or coming closer?*

Name: _____

Making shadows 1

You will need:

A torch, an object or toy, a screen or wall.

What to do

- Make a shadow.

- Try to make the shadow bigger.

- Try to make the shadow smaller.

- Try to make the shadow change shape.

- Draw a picture below to show how you made the shadow.

teddy bear

torch

screen

Cambridge Primary: Ready to Go Lessons for Science Stage 2 © Hodder & Stoughton Ltd 2013

Name: _____

Making shadows 2

1. How did you make the shadow bigger?

    _____

    _____

2. How did you make the shadow smaller?

    _____

    _____

3. Did you make the shadow change shape? yes / no

4. Draw the shadow that each object would make.

Object	Shadow

Playing with shadows

- Use first-hand experience. (2Ep2)
- Use a variety of ways to tell others what happened. (2Eo5)
- Explore how the Sun *appears* to move during the day and how shadows change. (2Pb1)

Resources

Outdoor space on a sunny day; chalk; a familiar storybook; torches or lamps; white cloth to make a screen; card; scissors; thin sticks (or pencils); sticky tape; photocopiable page 139; brass paper fasteners.

Starter

- Play some shadow games outside (see the lesson plan on page 23 for ideas).
- Reinforce ideas about shadows, for example they are always the same shape as the object that makes them, they can be made bigger or smaller depending on where the object or person is in relation to the light.

Main activities

- Read and share a favourite story with the class. This could be a traditional tale; choose one with several characters in it.
- Discuss the characters and their part in the story. Re-tell the story, inviting the learners to participate in repetitive phrases that define the main characters.
- Explain that you are all going to make a shadow puppet of one of the characters in the story and then, in groups, you will re-tell the story using the puppets.
- Set up the screen made from a cloth in a suitable space and appropriate height for the learners to perform behind. Demonstrate how to make a shadow puppet (you may want to have ready-prepared templates of characters for the learners to use). Cut it out and fix it to a wooden stick using sticky tape. Hold it close to the screen, using a torch for the light source. Show the learners that the closer the puppet is to the screen, the sharper the shadow will be.

- Give the learners time to make a puppet from the story. Make sure that you will have all the characters represented. Discuss that features are not important, but, for example, a king would need to have a crown on his head to show he is king.
- Copy photocopiable page 139 onto card and give it out to the learners who need extension. This gives instructions for how to make a moving crocodile. They might like to do this first, and then to try making their own moving shadow puppets for the story. Alternatively, they could make up their own story involving the moving crocodile.

Plenary

- Ask the learners in groups to perform the story as a shadow-puppet show.
- Evaluate each group's performance by asking the learners to comment on what they enjoyed particularly in each performance.

Success criteria

Ask the learners:

- What is the light source for our performance?
- Which character was easy to identify?
- Why didn't we need to draw faces on our puppets?
- How did the crocodile puppet move?

Ideas for differentiation

Support: Assist these learners with cutting skills and fixing the wooden stick to their puppets.

Extension: Challenge these learners to make a moving shadow puppet using photocopiable page 139.

Making a moving shadow puppet

You will need:

Scissors, sticky tape, two wooden sticks, a brass paper fastener.

What to do

- Cut out the crocodile.
- Ask an adult to make a hole for the brass fastener to go through.
- Join the body and tail using the brass fastener.
- Attach one wooden stick to the body and one to the tail.
- Make the crocodile move!

Testing shadows

- Collect evidence by making observations when trying to answer a science question. (2Ep1)
- Identify simple patterns and associations. (2Eo7)
- Explore how the Sun *appears* to move during the day and how shadows change. (2Pb1)

Resources

A sunny space outside; chalk; measuring tapes, metre sticks or a trundle wheel; photocopiable pages 141 and 142.

Starter

- Go outside and play some shadow games again. These are fun activities that reinforce the learners' understanding of what shadows are, how they are made and how they can be changed. For details of games, see the lesson plan on page 23.
- Draw around the outline of a learner's shadow with chalk. (You will come back to this at the end of the lesson.)
- Ask the learners to talk with talk partners about how shadows change during the course of a day. Listen to their responses.

Main activities

- Ask the learners for suggestions as to how they could find out exactly how shadows change during the course of the day.
- Discuss their responses and explain that in this lesson they will try to find that out for themselves. (Having drawn around the outline of a learner may prompt their thinking and reasoning.)

- Guide the learners in pairs into choosing an object (person or anything that makes a shadow) and designing a way to record how it changes. This could include drawing around the outline or taking photographs for later comparison. Allow the learners to pursue any reasonable requests. Some may want to measure the length of the shadow created each time. Ensure that they are confident in using the measuring equipment that is available.
- Give out photocopiable pages 141 and 142 and explain that they need to record on here to show how they did the test.

Plenary

- Draw around the same person at the end of the lesson and compare the shadows.
- Talk about how the shadow has changed in **size** during the course of the lesson (or day).
- Talk about how the shadow has changed its **position**.
- Compare this with what they found out from their tests.
- Explain that shadows are longer in the morning and late evening. At noon, shadows are short because the Sun is directly overhead.

Success criteria

Ask the learners:

- How did you set up the test?
- How did the shadow change in size?
- How has the position of the shadow changed?
- How long was the shadow at the start?
- How long was the shadow at the end?

Ideas for differentiation

Support: Work in a small group with these learners and help them to set up their own test. Assist them in completing photocopiable page 141.

Extension: Ask these learners to find out how shadows change in different seasons (for example the length of shadows at noon in summer compared to winter).

Name: _____

Testing shadows 1

List what you used here:

_____

_____

_____

_____

_____

Draw or write about what you did.

Name: _____

Testing shadows 2

1. Draw or write about what happened:

At the start	At the end

2. When was the shadow the shortest?

_____

3. When was the shadow the longest?

_____

Cambridge Primary: Ready to Go Lessons for Science Stage 2 © Hodder & Stoughton Ltd 2013

Investigating shadows 2

Learning objectives

- Take simple measurements. (2Eo4)
- Make comparisons. (2Eo6)
- Explore how the Sun *appears* to move during the day and how shadows change. (2Pb1)

Resources

A place in a sunny spot outside that can be separated off; plastic cones or poles; chalk; measuring equipment; graph paper; coloured paper; scissors; glue; flipchart and markers or whiteboard; computer; photocopiable page 144.

Starter

- Ask the learners to think back to the previous lesson and talk about what happened to the shadows that they tested (they changed in size and position).
- Ask the learners to describe how the shadows changed, for example got longer or shorter.
- Take the class outside and show them a marked-off area in a sunny spot. Explain that this is where they will collect shadow measurements in this lesson and throughout the day.

Main activities

- Position a cone or pole in the ground and measure the length of the shadow. Record the measurement on photocopiable page 144.
- Ask the learners to discuss with talk partners how long they predict the shadow will be one hour later. Listen to their responses, and organise them in pairs to take the measurement throughout the day at hourly intervals. (In mixed-ability pairs, the more-confident learner can do the measuring and the less-confident learner can record the measurement.) Ask them to record these on photocopiable page 144.
- Using an imaginary set of results, show the learners how to produce a bar chart from them. Model this on the flipchart or interactive whiteboard. Alternatively, produce a bar chart using computer software (if available).

- As the results come in throughout the day, add them to a blank bar chart. Discuss with the learners each time where to mark the length of the shadow on the bar chart. At the end of the activity, the bar chart can be used to find out the learners' level of understanding in interpreting data.

Plenary

- Look together as a class at the completed bar chart.
- Compare it with the bar chart drawn from imaginary results at the beginning of the lesson. Ask: *How are they similar?*
- Discuss the pattern that the bar chart produces and discuss why and when the shadow is shorter or longer.

Success criteria

Ask the learners:

- How long was the shadow at the start?
- How long was the shadow when it was the longest length?
- What time was the shadow shortest?
- Can you tell me why?
- How long did we do the test for?
- What is the difference between the longest and shortest shadow in length?

Ideas for differentiation

Support: Draw around the shadows on paper, then cut them out and use these to make a graph for classroom display by sticking them on graph paper and adding a title.

Extension: Use computer software to draw different types of graphs using the data obtained.

Name: _____

Investigating shadows

1. Record the length of the shadow every hour.

Time	Length of shadow

2. Draw a bar chart to show your results.

    Title: _____

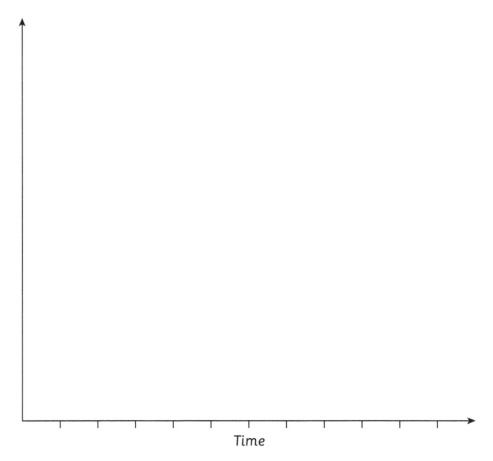

Time

Cambridge Primary: Ready to Go Lessons for Science Stage 2 © Hodder & Stoughton Ltd 2013

Make a sundial

Learning objectives

- Make and record observations. (2Eo3)
- Review and explain what happened. (2Eo9)
- Explore how the Sun *appears* to move during the day and how shadows change. (2Pb1)

Resources

Display the bar chart made by the less-able learners in the previous lesson; a pre-prepared model sundial (see photocopiable page 146); pencils; cotton reels; modelling clay or sticky tack; photocopiable page 146; books or internet access.

Starter

- Ask the learners who made the display bar chart in the previous lesson to show it to the rest of the class and explain what it tells them about how shadows change over time.
- Remind the learners that the shadow was shortest around the middle of the day. Ask: *What time of day was the shadow the longest?*

Main activities

- Show the learners the model of the shadow stick that you have prepared earlier. Explain that, in groups, they will each make one of these and use it to track the Sun's movement across the sky during the day. Show them how to do this.
- Alternatively, just use the model you have prepared for the whole class to use – but this will not be so much fun.
- Give out photocopiable page 146 to each group, as it has the instructions for making the shadow stick and a space for them to record what happens. Allow the groups to make these and choose a suitable place to leave them, where they will not be disturbed during the course of the experiment.

- Explain to the learners that at regular intervals, such as every hour, they will need to mark the position of the top of the shadow, using a cross or a dot. At the end of the experiment, they can join the dots to see the apparent path of the Sun throughout the day.

Plenary

- Ask the learners to describe the path that the Sun appears to make across the sky (a curve).
- Remind them that it is not the Sun that moves but the Earth, and that they will find out more about this in later lessons.
- Talk about what the learners know already about how the Earth moves.

Success criteria

Ask the learners:

- How did you make the shadow stick?
- What pattern did you see at the end?
- How does the Sun move across the sky?
- Does the Sun move?
- What else can you tell me about the Sun?

Ideas for differentiation

Support: Give these learners some sticky tack. Stick a piece onto a window to show the position of the Sun. Repeat during the day at hourly intervals instead of or as well as making the shadow stick.

Extension: Ask these learners to research Sun facts using secondary sources (books and the internet). Ask them to produce posters for display in the shape of the Sun.

Name: _____

Make a shadow stick

You will need:

A pencil, a cotton reel and a sunny place.

What to do

- Put the pencil in the hole in the cotton reel.

- Put it in one corner of the square below.

- Mark the top of the shadow with a cross every hour.

- Use a bigger piece of paper if you need to!

What does this tell you about how the Sun seems to move?

_____

Cambridge Primary: Ready to Go Lessons for Science Stage 2 © Hodder & Stoughton Ltd 2013

The Sun and Earth

Learning objectives

- Make and record observations. (2Eo3)
- Make comparisons. (2Eo6)
- Model how the spin of the Earth leads to day and night, e.g. with different sized balls and a torch. (2Pb2)

Resources

Globe (the Earth); a large inflatable ball (the Sun); a label 'Sun'; torch; photocopiable pages 148 and 149; internet access.

Starter

- Ask the learners to explain to you again how the Sun moves across the sky. (This is a very difficult concept for some young learners; some of them will think that the Sun literally rises and sets every day and night.)
- Recap on the fact that it is the Earth not the Sun that is moving.
- If possible, show a film clip of the Earth rotating on its axis.

Main activities

- Ask the learners to describe what they have seen.
- Explain that the Sun is at the centre of our solar system. Demonstrate this by asking one of the learners to hold the large, inflatable ball. If it helps, label this 'Sun'. Then, using the globe, orbit the Sun, that is, physically walk around the 'Sun', holding the globe, and explain that this is what makes a year ($365\frac{1}{4}$ days).
- Then concentrate on the Earth (globe) and demonstrate it revolving on its own axis. Show the learners the imaginary axis that goes from the North to South poles. Explain that at the same time as it is orbiting the Sun the Earth rotates on its axis. Again, this is a difficult concept. It will be much better understood by visual learners who are able to see this demonstration.

- Using a torch as the Sun, demonstrate the Sun as being still in the centre of our solar system and move the globe slowly on a single rotation to show day and night. It is sometimes helpful to place a marker on the surface of the globe, marking the place where the learners live.
- Give out photocopiable page 148 to the less-able learners and photocopiable page 149 to all the other learners. These pages require the learners to sort facts about the Sun and Earth, day and night.

Plenary

- Invite the learners in groups to show the rest of the class how day and night occur using the globe and the 'Sun'.
- Go over the answers to photocopiable pages 148 and 149.

Success criteria

Ask the learners:

- How does the Earth move?
- How long does a day and a night take?
- How long does it take for the Earth to orbit the Sun?

Ideas for differentiation

Support: Give these learners photocopiable page 148 and work through it with them.

Extension: Ask these learners to find out about any of the other seven planets besides Earth that orbit the Sun, that is, Mercury, Venus, Mars, Jupiter, Saturn, Uranus and Neptune.

Name: _____

The Sun and Earth

1. Use these words to help you complete the sentences below.

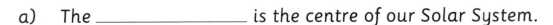

| axis | day | Earth | night | Sun | turns |

a) The _____ is the centre of our Solar System.

b) The Earth spins on its own _____.

c) It takes a year for the _____ to orbit the Sun.

d) Every 24 hours, the Earth _____ once. This gives us

   _____ and _____.

2. Add labels for day and night on the picture below.

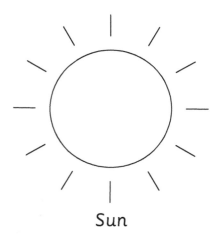

Sun

Earth

3. Write a list of things you do in the day and things you do at
    night-time.

Day   **Night**

_____   _____

_____   _____

_____   _____

_____   _____

Cambridge Primary: Ready to Go Lessons for Science Stage 2 © Hodder & Stoughton Ltd 2013

Name: _____

The Sun and Earth

1. Shade the diagram to show day and night on Earth.

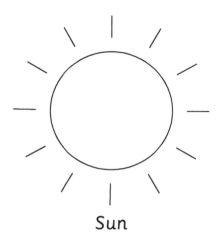

Sun

Earth

2. Join both parts of the sentences below together to make sense.
    Use a ruler to draw a line. The first one has been done for you.

The Earth	on its own axis.
The Sun	orbits the Sun.
The Earth spins	24 hours.
A day and a night take	stays still.

3. How many days is a year?

_____

4. What takes a year to happen?

_____

Using a sundial

- Take simple measurements. (2Eo4)
- Identify simple patterns and associations. (2Eo7)
- Explore how the Sun *appears* to move during the day and how shadows change. (2Pb1)

Watches or a clock; pictures (from books or the internet) of sundials; pieces of card; cotton reels; pencils; rulers; photocopiable pages 151 and 152.

Starter

- Ask the learners to think back to the lesson when they made shadow sticks (see page 145). Ask them to describe what they did and how they used the sticks. Show the shadow stick again.
- Ask the learners to think with talk partners about how people told the time before clocks and watches were invented. Listen to their responses.
- Show pictures of sundials from books or the internet.

Main activities

- Explain that the learners are going to set up another shadow stick in this lesson. They can do it in exactly the same way as they made the shadow stick on photocopiable page 146.
- This time, though, they are going to record differently, but in a similar way.
- Demonstrate and describe how to set it up:
  - If possible, make and set this up early in the school day. Otherwise, make them one day and set them up ready to use the day after. It needs to be set up and ready for the first line to be recorded on the hour.
  - Set up a shadow stick in a place that catches the Sun. Draw a line with a pencil and a ruler that passes through the centre of the pencil's shadow on the hour and write the time at the side of the line. Repeat this hourly throughout the day.

- Either make the shadow sticks in groups, or just make one for the whole class to use – but this will not be as meaningful to the learners.
- Remind the learners that the length of the shadow will change during the day. If any shadow lines are missed (lunch or break time), if the sundial is left in the exact same place, these lines can be drawn in another day.
- Once it has been marked with the hours, the sundial will show the time whenever the Sun is shining.

Plenary

- Look at the sundial when it has been marked off in hours.
- Discuss what the learners noticed about the shadow on each hour (discuss this as the recordings are being made).

Ask the learners:

- How does a sundial work?
- What does it need to work properly?
- Why are sundials not used very often today?
- How reliable is a sundial for helping to tell the time?

Support: Help these learners in making their group's recordings, if group sundials have been made. Assist them in completing photocopiable page 151.

Extension: Give these learners photocopiable page 152. This gives opportunities for trying to tell the time by reading sundials.

Name: _____

Sundials

Sundials were used to tell the time many years ago.

1.   Look at the picture and write a few sentences to say how a
     sundial works.

     _____

     _____

     _____

     _____

2.   Why are sundials not used much today?

     _____

     _____

3.   Where might you see an old sundial?

     _____

     _____

Name: _____

Sundials

Sundials were used long ago to tell the time.

What time do you think it is, using these sundials?
Look at the direction the Sun is shining each time.

(clock showing 12, sun above pointing down)	The time is _____ o'clock.
(clock showing 3, sun to left pointing right)	The time is _____ o'clock.

Cambridge Primary: Ready to Go Lessons for Science Stage 2 © Hodder & Stoughton Ltd 2013

Measuring shadows

- Use first-hand experience. (2Ep2)
- Take simple measurements. (2Eo4)
- Explore how the Sun *appears* to move during the day and how shadows change. (2Pb1)

Outdoor space; a tree, or other tall object that makes a shadow; flipchart and markers or interactive whiteboard; metre sticks or tape measures; photocopiable pages 154 and 155; calculators; height measurer (if available).

Starter

- Go outside and ask the learners to make the longest shadow they can, standing alone and using just their body.
- Then ask them to make the shortest shadow that they can, again just using their body.
- Discuss the shapes and techniques they use. Ask: *Which are the most effective? Why?*
- Repeat the activity, this time calling out numbers for the number of learners to get into a group and carry out the same activities.
- Do this for several different numbers in the groups, ending, if possible, with the whole class making the longest and shortest shadows that they can together in a group.

Main activities

- Ask the learners to walk with talk partners around inside a designated area and find the tallest shadow they can.
- Re-group as a class and share the learners' findings.
- Complete a table on the flipchart or interactive whiteboard of their discoveries. List the place, the object that made the shadow and its length.
- Ask them to estimate how tall the object is that is making a long shadow.
- Show them how to work out how tall the object is, using this information:
  - Measure the object's shadow to the nearest metre – ask some of the learners to help you.

- Measure the stick's shadow to the nearest decimetre (nearest 10 cm). (See photocopiable page 154.)
- Using a calculator, divide the length of the object's shadow by the length of the stick's shadow to give the height of the object.
- For example: if the object's shadow measures 12 m and the stick's shadow is 1.5 m, $12 \div 1.5 = 8$ m.
- Give out photocopiable page 154 and allow the learners to measure the length of the shadow they have chosen and the length of the shadow made by the stick. Provide plenty of support with the division sum.

Plenary

- Go around the groups, comparing predictions with calculated heights.
- Ask the learners who completed photocopiable page 155 (see 'Differentiation' section, below) to share their findings.

Ask the learners:

- How closely did predictions match calculated figures?
- Which is the tallest object?
- How tall is the tallest person in the more-able learners group, when measured using the shadow method?
- How tall is that learner measured with everyday measuring equipment?

Support: Assist these learners in using the measuring equipment. Supervise them using the calculator to find the answer.

Extension: Ask these learners to repeat the activity, measuring the shadow of the tallest learner in their group. Ask them to use photocopiable page 155 to record their calculations. Calculate the height from the shadow calculation. Then, compare this to the actual height measurement of the same learner. Ask: *How do the two measurements compare?*

Shadow calculations

Name: _____

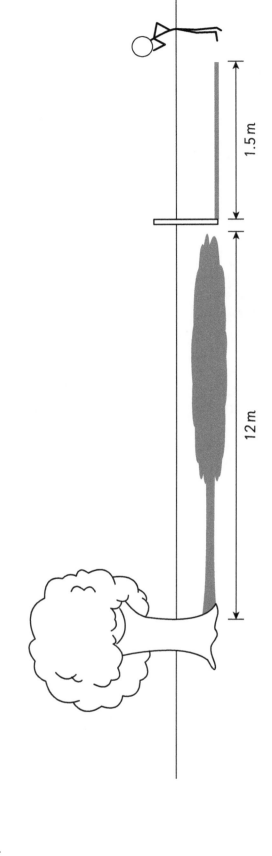

1.5 m

12 m

Complete this as your teacher works through the activity with you.

My object is a _____ .

Length of shadow = _____ m

Length of stick shadow = _____ m

Use a calculator:

_____ m ÷ _____ m = _____ m tall

Name: _____

Measuring height

1. Who is the tallest person in your group? _____

2. Draw a picture below to show how you measured the shadows.

3. Now check the height by doing the following sum:

    Length of person's shadow = _____ m

    Length of stick shadow = _____ m

    _____ m ÷ _____ m = _____ m tall

4. How tall is the person measured using a height measurer or metre stick? _____ m

5. Are the two measurements the same? yes / no

All about shadows

Starter

• Use this lesson as a revision session for the chapter before doing the unit assessment.

• Go outside and repeat some of the shadow activities again from some of the previous lessons in this unit, such as shadow-chasing games (see page 25) and making different sizes and shapes of shadow (see page 27). Ensure that the learners know and understand that shadows are formed when light is blocked, they change over time, but it is only their size and position that change – not the shape.

Main activities

• Set up a series of play activities around the classroom for the learners to revisit ideas for making shadows, for example a shadow-puppet theatre; shadow-measuring activities; shadow stick and sundials; and Earth and Sun (day and night) activities. Set out again as many activities as you can in the time and space available.

• Quickly explain to the learners that this lesson will include lots of activities about shadows that they have done before in this unit of work. Explain which particular activities have been set up and that they will have the opportunity to have a go at all the activities during this lesson.

• Organise the lesson so that all the learners have the opportunity to visit and do all the activities that you have set up. Make this a fun session.

• Circulate the different activities as the learners are working. This will give you an insight of their understanding as they interact with each other and with you as teacher. Make notes or record any significant findings, for example the learners who have a good grasp of a particular concept or no grasp at all. This can inform future planning.

Plenary

• Give each group of learners a wordsearch from photocopiable pages 157, 158 or 159. Give the less-able learners photocopiable page 157, the more-able learners photocopiable page 159 and all the other learners photocopiable page 158.

• Alternatively, the wordsearch activities could be a part of the Main lesson of activities, with a group at a time doing the wordsearch while the other groups carry on with the activities.

Name: _____

Shadows wordsearch

Read across or down to find these words.

o	d	t	u	r	n	E
c	a	x	i	s	l	a
d	y	s	o	h	p	r
l	S	y	e	a	r	t
i	u	e	f	d	u	h
g	n	b	l	o	c	k
h	a	l	i	w	e	x
t	n	i	g	h	t	l
s	o	u	r	c	e	m

Name: _____

Shadows wordsearch

Read across, down or up to find these words.

axis	block	day	Earth	light	night	
orbit	shadow	source	spin	Sun	year	

y	s	o	u	r	c	e
a	b	v	y	y	n	l
s	l	r	a	e	u	i
h	o	p	d	a	S	g
a	c	E	a	r	t	h
d	k	o	n	a	h	t
o	r	b	i	t	g	f
w	i	g	p	r	i	a
a	x	i	s	m	n	y

 Cambridge Primary: Ready to Go Lessons for Science Stage 2 © Hodder & Stoughton Ltd 2013

Name: _____

Shadows wordsearch

Read across, down, up or diagonally to find these words!

axis	block	day	Earth	light	night
orbit	rotate	shadow	source	Sun	year

l	i	g	h	t	s	b
a	S	u	n	i	i	l
w	b	y	s	b	x	o
o	E	d	o	r	a	c
d	f	a	u	o	n	k
a	c	y	r	e	i	l
h	f	a	c	t	g	s
s	e	r	e	q	h	p
y	r	o	t	a	t	e

Unit assessment

Questions to ask

- How long does it take for the Earth to spin once on its axis?
- How does the Earth move in space?
- Does the Sun move?
- What time of day are shadows shortest?
- How long does it take for the Earth to orbit the Sun?
- How many days are there in a year?

Summative assessment activities

Observe the learners while they participate in these activities. You will quickly be able to identify those who appear to be confident and those who may need additional support.

Earth in space

This activity allows the learners to demonstrate how much they remember about the Sun and Earth in space.

You will need:

A globe; a beachball or a large balloon – preferably yellow or orange (the Sun).

What to do

- In small groups, or with individuals, ask the learners to show you how the Earth moves, using the globe. (Spin it on its axis.)
- Introduce the item that you are using as the Sun. Ask one learner to be the Sun and hold it. Invite the other learners to be the Earth, using the globe to trace out the Earth's orbit.
- Record the names of those learners who know that the Sun does not move and that the Earth spins on its own axis.

Day and night

This activity allows the learners to show how day and night happen.

You will need:

A globe; a torch; a figure stuck on the globe at your country's location.

What to do

- With individuals, or in small groups, ask the learners to demonstrate to you how we get day and night.
- For those unable to do this, show them the Earth's rotation, using the torch as the Sun. Ask these learners to tell you when it is day and night for the figure attached to the globe.
- Record the names of those who clearly understand about how we get day and night. (This is a difficult concept.)

Written assessment

Distribute photocopiable page 161. The learners should work independently, or with the usual adult support they receive in class.

Name: _____

Day and night

1. Read the start of the sentence.

2. Draw a line to join it to the rest of the sentence.

The Sun	on its axis.
The Earth spins	take 24 hours.
The Earth orbits	365 days.
A day and night	does not move.
A year is	the Sun.

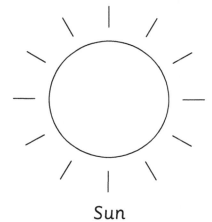

Sun

Earth

Animals around us

Learning objectives

● Use first-hand experience. (2Ep2)
● Make comparisons. (2Eo6)
● Identify similarities and differences between local environments and know about some of the ways in which these affect the animals and plants that are found there. (2Be1)

Resources

Flipchart and markers or interactive whiteboard; outdoor space with habitats for animals; gloves; photocopiable page 163.

Starter

- This unit builds on work covered in Stage 1, Unit 2A: 1.3 Living and growing and Unit 3B: 1.6 Growing plants.

- Explain to the learners that they will find out more about animals and plants around us, how important these things are to us and what they need to thrive.

- Ask the learners to think with talk partners about any animals that can be found living naturally locally. Ask: *What are these animals and where do they live?*

- List the learners' responses (or let them write) on the flipchart or interactive whiteboard.

- Use the information to construct a concept map (see page 132) and display this clearly for all the learners to refer to during this unit of work.

Main activities

- Ask the learners if they can remember what we call the place where an animal lives (a habitat – not a home). Think about and list any local habitats that they have looked at in previous units, such as in a pond, at the beach, in a tree, in a wall, under rocks, under leaves, in a log pile.

- Think about each habitat the learners mention in turn, considering what type of environment it is and which kinds of animals can be found there.

- Explain to the learners that in this lesson they will walk around outside and compare two different habitats. In pairs or small groups, they either need to choose for themselves or be directed by you to specific habitats.

- Give out photocopiable page 163 for the learners to complete as they compare the two habitats. Go outside so that the learners can explore. Remind them to leave the habitats exactly as they found them, for example do not leave stones over-turned or harm the animals.

- Make sure that the learners wear gloves and wash their hands after collecting and / or touching specimens.

Plenary

- Return to the classroom to share findings. Discuss and compare what the learners have seen.

- Talk about examples where there are many or few animals found and discuss possible reasons for this.

Success criteria

Ask the learners:

● Which habitats did you look at?
● Which creatures did you find there?
● What was your first habitat like?
● How was the second habitat different?
● Which habitat had the most creatures found in it?
● Which habitat had the least creatures in it?

Ideas for differentiation

Support: Working as a small group with these learners, pre-select two very different habitats for them to explore. Guide them in completing their findings in writing, where necessary.

Extension: Ask these learners to choose one particular animal that they have found and find out more about it.

Name: _____

Comparing animal habitats

Complete the table below for the two habitats you have found.

Habitat (for example pond)	Animal found (write or draw)	Describe the habitat (Hot or cold? Light or dark?)

Plants around us

- Use first-hand experience. (2Ep2)
- Make comparisons. (2Eo6)
- Identify similarities and differences between local environments and know about some of the ways in which these affect the animals and plants that are found there. (2Be1)

Photocopiable page 165; clipboards; digital cameras (if available); pictures (from books or the internet) of possible plants to be found.

Starter

- Recap on the previous lesson and think about the animals that were found in different habitats in the locality.
- Think about how each habitat is suitable for their needs – food, shelter, and so on.
- Ask the learners to share specific examples, such as a fish living in a pond (i.e. in water); it has gills and fins for swimming.
- Explain that in this lesson the learners are going to return to the same habitats that they visited last time to look at the plants they can find there.
- Ask the learners to discuss with talk partners if they can remember or predict any plants that they might expect to find in those habitats.
- Listen to their responses and tell them that they will have the opportunity to find out in this lesson.

Main activities

- Discuss rules for looking at plants, for example not to pull them up, pick flowers, or eat any fruits or berries. Emphasise the importance of good hygiene and the thorough washing of hands after the activity.

- In the same pairs or small groups, return to the same habitats visited last lesson. Ask the learners to think about any plants that they find there. *What is the plant called? Why is it useful to the animals that live in this habitat?*
- Give out photocopiable page 165 and explain that the learners need to use this page to record their findings. Describe and demonstrate or ensure that the learners are confident in using any cameras that are available for use.
- Return to the classroom or stay outside and visit each habitat for the learners to show the rest of the class the plants they have found.

Plenary

- Discuss the plants found, and try to name them. Use reference books or the internet back in the classroom, if necessary.
- Discuss the main purposes of plants for animals, for example as food or shelter.

Ask the learners:

- Which plant did you choose in this habitat?
- Do you know the name of this plant?
- What is it good for?
- How is this plant similar to plants other groups have found?
- What do animals need plants for?

Support: Assist these learners with taking photographs if necessary and finding out the names of plants that they do not know from reference books or the internet.

Extension: Ask these learners to make a poster showing all the different plants and habitats identified in the last two lessons. This can be an on-going activity throughout this unit.

Name: _____

Plants in their habitats

Find a plant in each habitat and complete the table below.

Draw the plant or add a photograph.	What is this plant called?	What is the plant good for? Shelter? Food? Anything else?

Plants growing in different places

Learning objectives

- Collect evidence by making observations when trying to answer a science question. (2Ep1)
- Make suggestions for collecting evidence. (2Eo1)
- Identify similarities and differences between local environments and know about some of the ways in which these affect the animals and plants that are found there. (2Be1)

Resources

Pictures (from books or the internet); photocopiable page 167; clipboards; magnifying glasses; digital cameras (if available); art materials – paints, brushes, paper, water; or collage materials – different papers, scissors, glue; or textiles materials – fabrics, threads, needles.

Starter

- Ask the learners to think back to the previous lesson and try to remember the names of any plants they found and identified.
- Explain that in this lesson you will ask them to find as many different plants as possible within a defined area.
- Ask them to think with talk partners about different places where they could look.
- Listen to their responses and either direct the learners to specific places, or leave them completely free to explore within a defined area.

Main activities

- Explain to the learners that they need to look at the plants they find and the conditions that they are growing in, for example: *Is it a wet or dry place? Is it in the Sun or in the shade? Is it near water? What shape are the leaves?*
- Give out photocopiable page 167 and explain to the learners that this is for them to record a particular plant in as much detail as possible. Ask them to work in small groups, where each learner records a different plant.

- Explain that they will use the information they have found, and reference books or pictures, to make a representation of a plant that they have found. This could be a close-up of the patterns found on a leaf (using magnifying glasses), or a representation of a flower, or of the whole plant.
- Provide opportunities for them to do a painting, collage or sewing of a plant – or just set up materials for them to produce one such item, that is, a painting **or** a collage **or** a sewn piece. This will depend on the materials available in school and your own expertise and confidence in these areas.

Plenary

- Look at the information shared by the learners and group the plants into those that grow in shady or sunny / wet or dry conditions.
- Discuss the characteristics of these plants and how they are adapted for survival, for example desert plants have thorns (to avoid being eaten by animals) and thick leaves and stems (to store water).

Success criteria

Ask the learners:

- Which plant did you find?
- What kind of habitat does it grow in?
- Why does it grow well there?

Ideas for differentiation

Support: Work with these learners in a small group. Pre-select the plants that you think they will be able to identify and make representations of confidently.

Extension: Ask these learners to make a detailed pencil drawing of the underside of a leaf, using a magnifying glass. Ask: *What can you see?*

Name: _____

Plant search

Use this page to collect ideas about a plant.

1. Name of plant: _____

2. Draw the plant in the box below. Remember the colours.
    Or you can add a photograph or make a collage.

3. Draw these parts of the plant after looking through a
    magnifying glass.

close-up of leaf

close-up of flower

Animals in their habitats

Learning objectives

- Use simple information sources. (2Ep3)
- Identify simple patterns and associations. (2Eo7)
- Identify similarities and differences between local environments and know about some of the ways in which these affect the animals and plants that are found there. (2Be1)

Resources

Pictures (from books or the internet) of a range of different animals; photocopiable pages 169 and 170.

Starter

- This lesson revisits what the learners know and have remembered about animal habitats from previous lessons and units of work.
- Ask the learners a variety of questions that are obviously incorrect, for example: *Do elephants live at the North Pole? Do kangaroos live in trees?* This should prompt them to reply with accurate answers that indicate their level of understanding.
- Show the learners a picture of an animal from another part of the world and ask them to tell you where that animal lives, for example a camel in the desert, a toucan in the rainforest.

Main activities

- Remind the learners that different animals and plants live in different places, called habitats. The learners have been looking at local habitats, but there are many different habitats around the world.
- The surroundings of a habitat create its environment – not its location. (This is difficult for some young learners to understand.) The surroundings, that is, the environment, affect what can live in a habitat. This includes things such as other animals and the air, weather and soil conditions. Animals and plants adapt over time to become more suited to the environmental conditions in their particular habitat.

- Ask the learners to think with talk partners about different environments around the world.
- Give photocopiable page 169 to the learners who need support and photocopiable page 170 to all the other learners. Explain that they have to match the animal to its habitat. Photocopiable page 170 also requires the learners to describe the environment. Show the learners any reference books or pictures available, or look on the internet, if available.

Plenary

- Go over the learners' responses to photocopiable page 169.
- Ask the other learners about the habitats and environments requested on photocopiable page 170.

Success criteria

Ask the learners:

- How is a camel suited to desert living?
- Describe the environment a polar bear lives in.
- Where does a blue whale live?
- Do chimpanzees and parrots live in similar environments?
- What is a habitat?
- What is an environment?

Ideas for differentiation

Support: Give these learners photocopiable page 169. Assist them with any research required – book or internet based.

Extension: Ask these learners to make a fact file about one of the animals discussed in today's lesson.

Name: _____

Animals in different habitats

Match the animal to its habitat by drawing a line.
The first one has been done for you.

camel

dolphin

polar bear

chimpanzee

ocean

desert

rainforest

North Pole

Name: _____

Animals in different habitats

Write about each animal's habitat and environment in the table below. The first one has been done for you.

Animal	Habitat	Environment
camel	desert	hot days, cold nights, not much water
penguin	_____	_____ _____ _____ _____ _____
blue whale	_____	_____ _____ _____ _____
parrot	_____	_____ _____ _____ _____

Cambridge Primary: Ready to Go Lessons for Science Stage 2 © Hodder & Stoughton Ltd 2013

Damage in the environment

Learning objectives

- Use simple information sources. (2Ep3)
- Talk about risks and how to avoid danger. (2Eo2)
- Understand ways to care for the environment. Secondary sources can be used. (2Be2)

Resources

Pictures (from books or the internet) of a variety of different worldwide environments; flipchart and markers or whiteboard; photocopiable pages 172 and 173.

Starter

- Remind the learners that an environment is the surroundings of the habitat where plants and animals live.
- Show the learners some pictures of different environments. Ask the learners to discuss with talk partners the types of animals, plants, water and vegetation found in that environment, for example a mountain – snow-capped summit, rocky, not much vegetation, animals could include mountain goats and eagles, and so on. Include other environments such as a rainforest, under the sea or a desert. Discuss these in a similar way, thinking about the animals, plants, water, temperature and soil conditions in each environment.

Main activities

- Explain that all around the world, environments are being damaged, destroyed and not cared for properly.
- As a class, compile a list of as many ways as you can think of in which the environment can be damaged. Act as scribe to write the list for the learners, or invite them to write the list. Record the list prominently on a flipchart or interactive whiteboard.
- This discussion will highlight if the learners are aware of the local, or local and global, situation. Begin the discussion from what they already know. Discuss local issues first, then gradually introduce issues of a more global nature over time.

- Discuss each of the learners' suggestions in turn, for example:
  - leaving lights on – wastes electricity or fuel
  - wasting water – dripping taps, leaking pipes, chemicals getting into water systems, such as from fields
  - rubbish left unattended – flies and rats spread disease and infection
  - ground clearance – burning trees for development
  - smoke pollution from industry
  - soil erosion on slopes.
- Give out photocopiable pages 172 and 173 and explain to the learners that they have to identify the things in the picture that are damaging to the environment. Give photocopiable page 172 to the learners who need support and photocopiable page 173 to all the other learners.

Plenary

- Ask the learners to highlight the evidence of environmental damage in the picture.
- Ask the learners who completed photocopiable page 173 to share their suggestions for improving that environment.

Success criteria

Ask the learners:

- How can humans damage the environment?
- Why is it important to have clean water?
- Where in the world do you think that this village could be?
- Why is litter or rubbish a bad thing to be around?
- What could we do in school to improve our environment?

Ideas for differentiation

Support: Give these learners photocopiable page 172 to complete.

Extension: Ask these learners to think of three ways in which the class could investigate around school if or where the environment is being damaged.

Name: _____

Damage in the environment

Circle as many things as you can find in this picture that are damaging the environment.

Cambridge Primary: Ready to Go Lessons for Science Stage 2 © Hodder & Stoughton Ltd 2013

Name: _____

Damage in the environment

1. Find three ways in which the environment is being damaged
    in this picture.

    a) _____

    b) _____

    c) _____

2. Suggest three ways in which this damage could be changed.

    a) _____

    b) _____

    c) _____

3. Write three different ways that you can show that you care
    for the environment.

    a) _____

    b) _____

    c) _____

Recycling

- Collect evidence by making observations when trying to answer a science question. (2Ep1)
- Use first-hand experience. (2Ep2)
- Understand ways to care for the environment. Secondary sources can be used. (2Be2)

Resources

Items suitable for recycling, e.g. aluminium drinks can, newspaper, empty plastic container, glass jar or bottle; picture (from a book or the internet) of a recycling logo; photocopiable pages 175 and 176; clipboards; computer package for producing graphs.

Starter

- Ask the learners to think back to the previous lesson and what things damage the environment.
- Ask them to think with talk partners about ways in which we can care for our school environment.
- Listen to their responses. Ensure you discuss ideas associated with recycling, but include references to saving water and not wasting energy.

Main activities

- Show some of the items available, such as metal, paper, plastic, glass, and ask if the learners know what we can do with these kinds of waste. Ensure that you use the word 'recycling'. If possible, look at the logo that indicates that a package or container is suitable for recycling. Ask the learners what this symbol means.
- Talk with the learners about things that they know are currently being recycled or can be recycled at school and at home, for example paper, metals, plastics, some fabrics.
- Explain that in this lesson they will conduct a survey around the school to find out where things are being wasted. The survey can be taken in pairs or small groups.

- Give out photocopiable page 175 to the learners who need support and photocopiable page 176 to all the other learners. Explain how to complete the page. This involves identifying places where waste is produced. Photocopiable page 176 also requires the learners to think about where the waste is created, what type of waste it is and whether it can be recycled or not. Remind the learners not to touch the waste, and stress the importance of hand-washing once back in class.

Plenary

- Go through the responses to photocopiable pages 175 and 176. Discuss the different types of waste and where they were found, for example paper in classroom waste bins, taps left running in cloakrooms.
- Think about which items can be recycled. Ask: *What effect does the waste found have on the school environment?*

Success criteria

Ask the learners:

- Did you find any paper waste?
- Where was it mostly?
- What kinds of waste things did you find that can be recycled?
- What recycling facilities do we have around school?
- What kind of things (if any) do you recycle at home?

Ideas for differentiation

Support: Give these learners photocopiable page 175. Work with them in a small group to conduct the survey.

Extension: Assist these learners with using their survey results to present them as a bar chart. Use a computer package to produce the bar chart, if possible.

Name: _____

Waste survey

1. Walk around school and find examples of waste and recycling.

2. Try to add an example for each of the materials listed in the table below.

Type of waste	Place found	Can it be recycled? (✓ or ✗)
paper		
plastic		
glass		
metal		
water		
electricity		

Name: _____

Waste survey

1. Walk around school and find examples of waste and recycling.

2. Complete the table below to show what you have found.

Place	Type of waste	Can it be recycled? (✓ or ✗)	Effect on environment?

3. Which type of waste did you find the most examples of?

_____

Cambridge Primary: Ready to Go Lessons for Science Stage 2 © Hodder & Stoughton Ltd 2013

Saving water

- Make suggestions for collecting evidence. (2Eo1)
- Take simple measurements. (2Eo4)
- Understand ways to care for the environment. Secondary sources can be used. (2Be2)

Bar chart from previous lesson; pictures (from books or the internet) of a dripping tap, a water butt, a bath or bowl full of water; flipchart and markers or whiteboard; photocopiable pages 178, 179 and 180.

Starter

- Ask the learners who produced the bar chart as an extension activity from the previous lesson to share it with the rest of the class.
- Discuss and interpret the results, asking such questions as: *What type of waste do we make most of in school? How do (or can) we recycle paper?*
- With talk partners, ask the learners to discuss how they could save water around school. Listen to their suggestions and ideas.
- Show the pictures available of the examples of familiar items that use or contain water. Discuss them in turn, thinking about if these things are ways in which we waste water, or are ways in which we can save or recycle water.

Main activities

- Explain that in this lesson the learners will find out how and where water is being wasted around school and then suggest ways to save and reuse it.
- Ask the learners to recall with talk partners examples of water being wasted around school. List the learners' responses on the flipchart or interactive whiteboard, and display it prominently.

- In small groups, ask the learners who need support: *What can we use recycled water for?* Ask all the other groups of learners: *How can we find out how much water is being wasted?*
- Give photocopiable page 178 to the learners who need support and explain that they need to identify what the water in the pictures can be used for. Give photocopiable pages 179 and 180 to all the other learners to record their measurements of the amount of water wastage.

Plenary

- Discuss the learners' responses to the photocopiable pages they have completed in the lesson. Discuss their ideas for using waste water from use at home.
- Talk about whether the learners think that the amount of water wasted is worth collecting.

Ask the learners:

- How did you measure the amount of water?
- How long do you think it might take to fill a bowl or bucket with drops of water?
- What can we use water collected in a water butt for?
- What could we do with water from a bath when we empty it?
- Does anyone know what can be done to fix a dripping tap?

Support: Give these learners photocopiable page 178 to complete.

Extension: Ask these learners to design and make a poster to encourage turning taps off, and display them in appropriate places as reminders around school.

Name: _____

Recycled water

1. Look at the pictures and say where the water that has been collected has come from.

Water container	Where has the water in it come from?

2. Give one answer for each of these questions:

a) Water from the water butt can be used for _____

b) Water in a bath can be used for _____

c) Water from a tap can be used for _____

Cambridge Primary: Ready to Go Lessons for Science Stage 2 © Hodder & Stoughton Ltd 2013

Name: _____

How much water are we wasting?

How could you find out how much water a dripping tap wastes?

_____

_____

_____

Now try to find out by measuring drips!

You will need:

a plastic cup or beaker

a dropper

water

What to do

- Use the dropper to drip water into the cup or beaker.

- Drip drops slowly, like a dripping tap, for one minute.

What happened?

We dripped _____ drops in one minute.

1. What could you do with this amount of water, instead of throwing it away?

    _____

2. What could you do with the amount of water you could collect in an hour?

    _____

Name: _____

How much water are we wasting?

1. If you counted the number of drops, do this calculation using a calculator.

    number of drops = _____

    Every hour we waste:

    number of drops _____ × 60 = _____ drops

2. Ask your teacher to pour this amount of water into a measuring container.

3. There are 24 hours in a day. So every day we waste (use a calculator):

    number of drops collected in an hour _____ × 24 =

    _____ drops

4. Now ask your teacher to pour more water into the container to show how much water is wasted every day in this way.

5. What could you use this wasted water for?

    _____

    _____

    _____

 Cambridge Primary: Ready to Go Lessons for Science Stage 2 © Hodder & Stoughton Ltd 2013

Litter

Learning objectives

- Use simple information sources. (2Ep3)
- Talk about risks and how to avoid danger. (2Eo2)
- Understand ways to care for the environment. Secondary sources can be used. (2Be2)

Resources

Paper; art materials; photocopiable pages 182 and 183; plastic gloves; pictures (from books or the internet) of litter-strewn places; a full classroom rubbish bin; secondary sources – books and internet access.

Starter

- Ask the learners if they can remember doing the survey on waste around school. Explain that in this lesson they will think about the different kinds of rubbish thrown in the bin every day during lessons.
- Prepare a typical classroom litter bin, containing items of a typical school day's rubbish, for example pencil shavings, paper, empty plastic water bottles, paper towels. Ensure there is nothing dangerous in the rubbish.
- Wearing plastic gloves, remove items of litter from the bin, or ask a learner to come forward and tell the rest of the class one thing that they can see in the bin. Talk about whether any of these things could be recycled.

Main activities

- Discuss why you are wearing plastic gloves for this demonstration. Talk about risks such as germs being spread from used tissues, possible dangers of broken glass and therefore cuts, and so on.
- Explain the importance of not touching litter in a bin for these reasons.
- With talk partners, ask the learners to think about public places in the local area. Ask: *Why is litter a danger outside in such places?* Listen to their responses (for example it creates bad smells, it can cause diseases to spread). Ensure that you discuss the dangers that litter creates for animals as well as for humans.

- Hold a class discussion on the following question: *Is litter a problem for us around school?* Invite as many learners as possible to contribute. Ask them to give reasons, if they can, and to suggest what could be done to improve the situation.
- Compile a list of class rules about how, in class, you could care for the environment, for example recycle different types of rubbish, don't drop litter, turn off lights when not in the room, make sure that taps are fully turned off.
- Think about ways in which the class could encourage similar behaviour around school.
- Give out photocopiable page 182 to the learners who need support and photocopiable page 183 to all the other learners. These activities help to summarise their understanding of some of the ideas around caring for the environment.

Plenary

- Emphasise the importance of saving electricity and water and reducing waste as much as possible.
- Discuss ways to improve recycling facilities in and around school.

Success criteria

Ask the learners:

- Which materials do we recycle in school?
- What other materials can be recycled?
- How can we save electricity?
- What can we do to save water?
- Why is litter a problem?

Ideas for differentiation

Support: Give these learners photocopiable page 182 to complete.

Extension: Ask these learners to prepare a presentation about caring for the environment to share in an assembly in school.

Name: _____

Caring for the environment

We can show that we care for the environment when we try not to waste things.

1. Draw and label three different things that can be recycled.

a)	b)	c)

2. One way we can save electricity is by

_____

_____

_____

3. One way we can save water is by

_____

_____

_____

Cambridge Primary: Ready to Go Lessons for Science Stage 2 © Hodder & Stoughton Ltd 2013

Name: _____

Caring for the environment

Use these words to complete the sentences below. Use each word only once.

environment   glass   lights   metal
plastic   recycle   rubbish   tap

1. We show that we care for the _ _ _ _ _ _ _ _ _ _ _
   when we try not to waste things.

2. We can _ _ _ _ _ _ _ things like paper and

   _ _ _ _ _ _ _.

3. We can also recycle _ _ _ _ _ and _ _ _ _ _.

4. We can save electricity by switching off _ _ _ _ _ _.

5. We can save water by turning off a dripping _ _ _.

6. To keep our environment tidy, we can put _ _ _ _ _ _ _
   in the bin.

How materials decay

Learning objectives

- Make and record observations. (2Eo3)
- Review and explain what happened. (2Eo9)
- Understand ways to care for the environment. Secondary sources can be used. (2Be2)

Resources

Classroom litter bin; aluminium foil; a core from a piece of fruit; cardboard; an empty sweet packet; leaves; soil; a piece of citrus peel; trays; plastic gloves; digging implements; photocopiable pages 185, 186 and 187; digital cameras (if available).

Starter

- Pre-prepare a classroom litter bin containing the kinds of items listed in the resources section above. Ensure there is nothing dangerous in the rubbish. Show the learners what rubbish is in the bin.

- Ask them to discuss with talk partners what would happen if each type of rubbish was simply thrown on the ground. Natural things would decay naturally and be absorbed into the soil. Paper products will disintegrate over time, but metals and some plastics (which are not biodegradable) will remain in the soil.

- Talk about dangers to animals, plants and humans that discarded litter can cause.

Main activities

- Explain to the learners how waste is processed locally. It might be possible to arrange a visit to a recycling centre or to invite a representative speaker into school to tell the learners more about recycling.

- Give the learners in pairs or small groups a piece of litter from the bin used in the Starter activity. Tell them that they are going to bury the item and observe it over time to see how well or badly it decays. They need to decide where they would like to bury it – somewhere where it will not be disturbed. (Alternatively, bring some large pots of soil into the classroom for the learners to bury the litter in.)

- Give photocopiable page 185 to all the learners except those who need extension to complete during this activity. Give photocopiable pages 186 and 187 to the learners who need extended tasks. This activity investigates conditions affecting decay. Give each learner in the group a different type of litter. This will allow for comparisons between the different rates of decay to be made at the end of the experiment. Remind the learners to look but not to touch any decaying materials.

Plenary

- Discuss what is happening to the rubbish week on week.

- Explain about natural decay.

- Predict what will happen to each type of rubbish eventually. Talk about how rubbish that does not decay is treated in your country.

Success criteria

Ask the learners:

- Which decayed the most?
- Is there anything that has not decayed at all?
- Which conditions helped the rubbish to decay most quickly?
- What happens to rubbish that does not decay?
- How did you make the test fair?

Ideas for differentiation

Support: Organise these learners into mixed-ability groups to carry out this activity.

Extension: Give these learners photocopiable pages 186 and 187 and ask them to set up an investigation into rates of decay in different conditions.

Name: _____

How materials decay

1. We buried _____ in the soil.

2. Complete the table below to describe what happened.
   You can write, draw or stick on photographs.

Week 1	Week 2	Week 3	Week 4

3. Compare your results with the other groups.

4. Complete the list to show what happened.

   | Decayed most: | _____ |

   _____

   _____

   _____

   | Did not decay: | _____ |

Name: _____

How materials decay

You will need:

Four plastic cups, four pieces of litter that are all the same such as a fruit core, soil, water.

What to do

- Put some soil in each cup.

- Bury a piece of litter in each cup.

- Label the cups A, B, C, D.

- Add water to cups B and D.

- Put cups A and B in a warm place.

- Put cups C and D in a cool place.

- Dig up the litter each week to see what has happened to it.

1. We buried _____ in the soil.

2. Complete the table below to describe what happened.
    You can write, draw or stick on photographs.

Week 1	Week 2	Week 3	Week 4

Cambridge Primary: Ready to Go Lessons for Science Stage 2 © Hodder & Stoughton Ltd 2013

Name: _____

Comparing results

1. Compare your results with the other learners in your group.

Litter	Decay or not? (✓ or ✗)

2. Which conditions caused most decay?

   Circle the correct answer.

   | cold and dry   cold and wet   warm and dry   warm and wet |

3. Why do you think this is?

   _____

   _____

Weather observations

Starter

- Introduce the idea that weather is a particular factor that influences our environment, or indeed any environment in the world.
- Ask the learners how they would describe the weather today. Ask: *What was it like yesterday? What might it be like tomorrow?*
- If possible, watch a weather report on television. This could be a general news report, or one taken from a children's television bulletin.
- Ask the learners to list with talk partners any weather words used in the weather report.
- Compile a class list of weather words and discuss what each term means. Display the list for the learners to refer to during the lesson.

Main activities

- Explain to the learners that they are going to observe and record the weather conditions during the next week. Have a large chart prepared, similar to the one provided on photocopiable page 189, or use an educational resource that you may already have for classroom use. These are widely available from educational suppliers.
- Show and demonstrate how to read the temperature on a thermometer. Show the learners how to write temperature in °C. Allow them to use thermometers to read temperatures in the classroom and outside.

- Show and demonstrate how to use the anemometer, which measures wind speed, and rain gauge, if available. Allow the learners time to use this equipment.
- Complete the class weather chart for today. Give out photocopiable page 189 for the learners to complete. Discuss the use of weather symbols and agree on some for the learners to use on their weather chart.
- Talk about the seasons and how the weather changes throughout the year in your country. Give out photocopiable page 190 for the learners to draw weather pictures for each season.

Plenary

- Summarise what the weather is like in your country through the seasons.
- Discuss what the weather is like today.
- Predict what the weather might be like tomorrow.
- Think about how people in your country adapt to living in such weather conditions.

Name: _____

Weather chart

1. Complete the weather chart below for five days.

2. You can write or draw weather symbols like these in the last row.

Day	1	2	3	4	5
temperature in °C					
wind speed					
cloud cover					
What was the weather like today?					

Name: _____

Weather through the year

1. What are the seasons of the year?

   S _ _ _ _ _ _   S _ _ _ _ _ _   A _ _ _ _ _ _   W _ _ _ _ _ _

2. Draw a picture to show what the weather is like in each season.

Cambridge Primary: Ready to Go Lessons for Science Stage 2 © Hodder & Stoughton Ltd 2013

Unit assessment

Summative assessment activities

Observe the learners while they participate in these activities. You will quickly be able to identify those who appear to be confident and those who may need additional support.

Caring for the environment

This activity involves a discussion about recycling and ways to care for the environment.

You will need:

A quiet corner to work in.

What to do

- Working on an individual basis, ask the learners in turn to tell you:
  - one thing that can be recycled
  - two ways to recycle water
  - three things we can do to show that we care for the environment.
- Record or take note of their responses.
- This discussion will summarise how much they know about caring for the environment in and around school.

Weather observations

This allows the learners to interpret weather data.

You will need:

A weather chart identical to that being used in class, with a different set of recorded weather results on it.

What to do

- Working with a small group, interrogate the weather data on the chart you have prepared.
- Ask the learners which were the highest and lowest temperatures recorded. Ask questions about wind-speed (if recorded).
- Include each learner, directing questions to each at their own level of understanding.
- Explain and discuss any misunderstandings as they arise.
- Their responses will indicate to you how much they know and understand about interpreting weather data.

Name: _____

Plants and animals around us

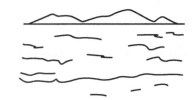

1. Fill in the table below by naming an animal and a plant that can be found in each habitat.

2. Complete the table by adding two habitats that you have looked at.

Habitat	Animal	Plant
pond		
desert		
in the sea		

Cambridge Primary: Ready to Go Lessons for Science Stage 2 © Hodder & Stoughton Ltd 2013